Carl Sagan's Universe

Carl Sagan's many contributions to science and society have been profound and far-reaching, influencing millions of people around the world. He carried out significant research in planetary science, was closely associated with the U.S. space program, created the highly acclaimed television series, *Cosmos*, and was the Pulitzer Prize-winning author of many best-selling popular science books. *Carl Sagan's Universe* is a fascinating and beautifully illustrated collection of articles by a distinguished team of authors and covers the many fields of science, education, policy making, and related areas in which Sagan worked.

The book is divided into four sections, the first two of which provide an absorbing overview of the U.S. space program (as well as a complementary account of the Russian program) and of the history and current status of the search for extraterrestrial life. The final two sections deal with the importance of science education in the successful development of a technological society and of the shaping of science policy in tackling the problems facing us today. Also included is a separate chapter by Sagan himself, discussing the place and role of our planet and mankind in the universe.

Written in honor of Carl Sagan's many achievements, this book will fascinate and reward anyone interested in planetary science and exploration, the search for extraterrestrial life, or the role of science in the modern world.

Carl Sagan's Universe

Edited by

Yervant Terzian
Cornell University

Elizabeth Bilson
Cornell University

CAMBRIDGE
UNIVERSITY PRESS

PUBLISHED BY THE PRESS SYNDICATE OF THE UNIVERSITY OF CAMBRIDGE
The Pitt Building, Trumpington Street, Cambridge CB2 1RP, United Kingdom

CAMBRIDGE UNIVERSITY PRESS
The Edinburgh Building, Cambridge CB2 2RU, United Kingdom
40 West 20th Street, New York, NY 10011-4211, USA
10 Stamford Road, Oakleigh, Melbourne 3166, Australia

© Cambridge University Press 1997

First published 1997

Printed in the United States of America

Typeset in Melior and Eurostile

Library of Congress Cataloging-in-Publication Data

Carl Sagan's universe / edited by Yervant Terzian, Elizabeth Bilson.
p.   cm.
ISBN 0-521-57286-X. — ISBN 0-521-57603-2 (pbk.)
1. Life on other planets.   2. Science news.   3. Science – Social
aspects.   4. Sagan, Carl, 1934–1996.   I. Terzian, Yervant, 1939–
II. Bilson, Elizabeth M.
QB54.C37   1997
500 – dc21   96-40511
                                                             CIP

*A catalog record for this book is available from
the British Library*

ISBN   0 521 57286 X hardback
ISBN   0 521 57603 2 paperback

Contents

List of Contributors

BILL G. ALDRIDGE
Director, Science Education Solutions
  Vice President, Airborne Research
  and Services

WALTER ANDERSON
Editor, Parade Publications

GEORGI ARBATOV
Director Emeritus and Chairman
  of the Governing Board, Institute
  of U.S. and Canadian Studies
  Russian Academy of Sciences

ELIZABETH M. BILSON
Administrative Director, Center
  for Radiophysics and Space
  Research, Cornell University
  Editor of this volume

JOAN B. CAMPBELL
General Secretary, National Council
  of the Churches of Christ

CHRISTOPHER F. CHYBA
Assistant Professor, Department of
  Planetary Sciences, The University
  of Arizona

FRANK D. DRAKE
Professor of Astronomy, University of
  California at Santa Cruz

ANN DRUYAN
Secretary, Federation of American
  Scientists

RICHARD L. GARWIN
IBM Fellow Emeritus
IBM Research Division

JAMES HANSEN
Director, NASA Goddard Institute
  for Space Studies

PAUL HOROWITZ
Professor of Physics
Harvard University

WESLEY T. HUNTRESS, JR.
Associate Administrator for Space
  Science, NASA Headquarters

JON LOMBERG
Senior Advisor
The Planetary Society

DAVID MORRISON
Chief, Space Science Division
  NASA Ames Research Center

PHILIP MORRISON
University Professor Emeritus
  Massachusetts Institute
  of Technology

BRUCE MURRAY
Professor of Planetary Sciences
  Division of Geological & Planetary
  Sciences, California Institute
  of Technology

FRANK PRESS
Senior Fellow, Carnegie Institution
  of Washington

JAMES RANDI
Plantation, Florida

FRANK H. T. RHODES
President Emeritus
  Cornell University

CARL SAGAN
David Duncan Professor
  of Astronomy and Director
  Laboratory for Planetary Studies
  Cornell University

ROALD SAGDEEV
Professor of Physics and Director
  East–West Science Center
  University of Maryland

EDWARD C. STONE
Director, Jet Propulsion Laboratory
  California Institute
  of Technology/JPL

YERVANT TERZIAN
James A. Weeks Professor of Physical
  Sciences; Chairman, Department of
  Astronomy, Cornell University
  Editor of this volume

KIP S. THORNE
Richard Feynman Professor and
  Professor of Theoretical Physics
  California Institute of Technology

OWEN B. TOON
Senior Scientist, Earth Science
  Division, NASA Ames
  Research Center

RICHARD P. TURCO
Professor of Atmospheric Sciences
  Department of Atmospheric
  Physics and Institute of Geophysics
  and Planetary Physics, University
  of California, Los Angeles

Preface

When Carl Sagan came to Cornell in 1968 he was young, brilliant, and ambitious; in this respect not so different from other new faculty members. But Sagan had uncommon vision and well-defined purpose. He was fascinated by science and by astronomy in particular, and he believed that key questions concerning the origins of life and the existence of life elsewhere in the universe could be confronted by rational thinking combined with astute research and observation. He was further convinced that what he knew and believed, and what he hoped to discover, had to be effectively communicated to the public policy makers and indeed to the general public at large. He recognized that in a technological society (or in any advanced society, for that matter) science is critical for informed decision making.

For nearly three decades we watched Carl Sagan pursue his vision with great dedication and spectacular success. He played a leading role in the American space program since its inception. He briefed the Apollo astronauts before their flights to the Moon and was an experimenter on the Mariner, Viking, Voyager, and Galileo expeditions to the planets. He helped solve the mysteries of the high temperature of Venus in terms of a massive greenhouse effect; he explained that the seasonal changes on Mars were caused by windblown dust; and he showed that the reddish haze of Titan was due to organic molecules in its atmosphere. He was a consultant and adviser as well as an important spokesperson for the National Aeronautics and Space Administration (NASA) and the entire scientific community at congressional hearings and in the press. He brought public attention to extremely important environmental and other issues, such as the Nuclear Winter. He was one of the key scientists who organized and inspired programs in the search of extraterrestrial intelligence.

Sagan became a best-selling author the world over of books that popularize science and its significance for mankind. In 1978 he received the Pulitzer Prize for *The Dragons of Eden*. In 1980 he presented the Public Television Series *Cosmos*, which was seen by 500 million viewers in sixty countries. This thirteen-part series broke all previous records in terms of viewers and had a great impact on people everywhere. It was a magnificent perusal of the birth and development of life, civilization, and science on Earth. The ensuing book, *Cosmos*, was on *The New York Times* bestseller list for seventy weeks and had forty-two printings in the American edition, plus thirty-one foreign editions.

Sagan was the best-known and most popular science writer and educator in this century. At Cornell a Sagan lecture filled any auditorium to capacity; there was fierce competition among students to register in his limited-enrollment classes. Many of the most productive planetary scientists working today were his former students and associates. He was a much sought after lecturer around the globe, not only because there was such interest in the subjects he discussed but also because of his extraordinary talents as a public speaker who could reach, educate, and indeed entertain any audience on Earth.

In 1980, Sagan and Bruce Murray formed The Planetary Society, dedicated to the exploration of our Solar System, the search for planets around other stars, and the quest for extraterrestrial life and intelligence in the universe; the Society, with Sagan as its first President, presently counts more than 100,000 members worldwide. What Sagan wrote about its goals best expresses his aspirations and personal philosophy: "to discover and explore new worlds, and to seek our counterparts in the depths of space – these are objectives of mythic proportions. They are now in the realm of sober scientific reality because of the enormous technological strides made in the last two decades. Pursuing these endeavors for the benefit of the human species is a mark of our dedication to a hopeful future."

At sixty, Carl Sagan still remained the same relentless worker and dreamer he was as a young scientist. His books followed one another in rapid succession, and he was getting ready to produce a motion picture based on his novel *Contact*. In his laboratories experiments were being conducted, simulating the atmosphere of Jupiter, the conditions on Titan, and still others to learn more about the origin of life on our own planet.

To celebrate Carl Sagan's sixtieth birthday, Cornell University organized a symposium dedicated to his work. This meeting took place in October 1994 at the Cornell campus and was attended by more than 300 scientists, educators, friends, and family from around the world. The papers presented in this volume were delivered in his honor during the symposium. The four general subjects, I. Planetary Exploration; II. Life in the Cosmos; III. Science Education; and IV. Science, Environment, and Public Policy, were discussed by an array of distinguished speakers and demonstrate Carl Sagan's interests and involvement during the last few decades.

Carl Sagan's work has inspired and motivated countless young people around the whole Earth to pursue the sciences. During the symposium banquet, a young student from Niamey, Niger, Hamadou Seini, related how Sagan's influence made him organize the "Carl Sagan Astronomy Club" in Niamey, and a young Cornell freshman, Baquera Haidri, expressed her attraction to science and Cornell due to Sagan's work, *Contact*, with the words, "... In her hand, clenched tightly, was a tattered old paperback, yet within its pages was the most valuable story she had ever read – a tale of adventurers called scientists, of struggles and victories, of life, and of an unimaginable wonderful journey to a place not only without us all, but within us as well. And it was as if suddenly this young girl, longing all her life to touch those pinpricks of infinity, had made *Contact*."

The Sagan symposium, whose proceedings appear in this volume, was sponsored by the Department of Astronomy at Cornell University, the New Millennium Committee of The Planetary Society, and *PARADE* magazine. We would like to thank Andrea Barnett, Sharon Falletta, and Laurel Parker for their cooperation and assistance, and Ann Druyan, Peter Gierasch, and Ed Salpeter for their invaluable advice. We are grateful to Mary Roth for her expert transcription of the proceedings and her valuable help in editing.

Several months after Sagan's sixtieth birthday, he was diagnosed with a rare disease, myelodysplasia. He fought the illness with never waning courage and optimism for nearly two years. On December 20, 1996 he died of pneumonia at the Fred Hutchinson Cancer Research Center in Seattle, where he had received a bone marrow transplant and a number of follow-up treatments.

Carl Sagan was buried in Ithaca on December 23, 1996. He was remembered and mourned around the entire globe. His family, friends, colleagues and all those he touched by his writings, lectures, speeches and television programs will deeply miss him for a long time to come.

Yervant Terzian
Elizabeth M. Bilson

January, 1997
Ithaca, NY

PLANETARY EXPLORATION

1

On the Occasion of Carl Sagan's 60th Birthday

WESLEY T. HUNTRESS, JR.

NASA Headquarters, Washington, D.C.

Carl Sagan is sixty years old, and the Space Age is only thirty-seven years old. Carl is so identified with the Space Age that it is hard to believe that he wasn't born with it. Over those thirty-seven years of the Space Age, Carl has been the single most-recognized science missionary bringing the ideas, excitement, and adventure of space exploration to the general public. One thing that Carl has done so well in his career is to make both scientist and layman *think*, and in particular to think about science and space exploration in a much larger societal and historical context. He has caused us to consider what it is about space exploration that is so fascinating to human beings and why an investment in space is so important for our future.

Over his long scientific, literary, and public career, I think that Carl's greatest accomplishment may be that he has become for many Americans an icon for modern science. For many on this planet, Carl is the personification of space science and exploration. He has reached millions of people with his articles, television appearances, and books. He has explained science and its importance to many an audience. Throughout his work, Carl has carried the message that the communication of science is crucial to its success. Carl realized long ago that scientists have a responsibility to participate in society. Scientists can no longer remain safely in their laboratories and offices, divorced from the rest of the world. Carl recognized early in his career that scientific discoveries have value only if they are shared. Carl understood more than most scientists that science can continue to prosper only if the public can participate in its excitement and thereby support its continuation. There have been those in the science community who have not fully appreciated the value that Carl's approach has brought to the scientific enterprise. The truth is that we need a lot more like him.

The chapters of this book clearly reflect the role Carl has played in promoting science and the scientific process. They start with a section

3

on "Planetary Exploration," the field in which he has been involved as a working scientist for most of his scientific career, including major roles in the Viking and Voyager missions. Then they move to "Life in the Cosmos," an area of science with which Carl is heavily identified (including his famous PBS series, *Cosmos*). There will be a chapter by Carl on "The Age of Exploration," allowing him to put science into the larger context of history, something he has also done many times and always to our great edification. There are two more sections, one on "Science Education" and one on "Science, Environment, and Public Policy." These are areas where science interacts with the larger world within which it moves. As I have mentioned, Carl has been a leader in these areas. In addition, some of Carl's greatest successes to date have involved separating science from pseudoscience, another topic that will be treated in this book.

I have always found a lot of inspiration in what Carl has to say. I've grown up in the business of space exploration listening to and absorbing Carl's thoughts on our enterprise. So in what follows are some of my own personal thoughts on what the space exploration enterprise is all about, and about a vision for the next century. My own vision may not be unique, but it contains the thoughts of a fellow traveler with Carl on the mind paths leading outward from planet Earth.

So what is it all about, this idea of space exploration? We like to talk about vision these days – vision of a future. It is only natural that as we stand near the end of this century today, we try to conjure a vision of the future as we gaze over the horizon into the next century. To me, the idea of future vision is to peer outward to where the human body and mind have not yet traveled. As an example, imagine an early human tribe 20,000 years ago confined to its small hunting/gathering territory. At some point, a particularly curious individual travels beyond the tribe's visible, territorial horizon and returns with new discoveries and knowledge of what lies beyond. Immediately, the tribe's resource and intellectual base has expanded. Their quality of life and chance of survival have been significantly enhanced. This is the nature of humanity: to explore beyond our horizons, whether they be territorial, sociological, physiological, scientific, or technological. Humanity explores in order to discover, discovers in order to gain new knowledge, and gains new knowledge in order to enhance the quality of life for itself.

It has been by looking beyond current horizons that human civilization advances. Cultures that turn in on themselves and confine themselves to their current horizons ultimately do not advance. So if you scan today for the most expansive and challenging of human horizons, certainly one of those is space. Space is not just a territorial horizon. Space is also an intellectual and scientific horizon, defining our understanding of the universe in which we live. Space is a technological horizon, driving development of sensors and instruments, information systems, computational systems, communication systems, navigation, automation, robotics, and a number of technologies that can improve the quality of life on this planet.

Space is a sociological horizon, challenging our concept of humanity and its role in the universe. Are we a species bound to earth, or should we expand off-planet? To me, the answer is self-evident. Does an understanding of what is beyond our planetary horizon help us to become more unified as a global society within its confines? I think it does. Just remember the worldwide reaction to the "Blue Marble" picture of our spaceship Earth taken from the first Apollo lunar session. Are we at all affected by what happens out there in that seemingly never-changing emptiness of space? Of course we are. Our solar system and universe are not the static, benign, predictable, clockwork environments that our never-changing view of the night sky would lead us to believe. The comet Shoemaker–Levy 9 impacts on Jupiter should have put all that to rest!

Space can be an economic horizon, because thirty-five years of planetary exploration has shown that there are resources in space. These resources could provide sustenance for the space-faring enterprise in the future and possibly provide economic benefit for people on earth as well. The value of space as an economic horizon remains to be proven, but at least today we can imagine with more credibility such ideas as imported power and materiel from space, communication services, and even entertainment and tourism. Consideration of the utilization of in situ resources in space for the production of fuel, oxygen, and other consumables has become a real feature of our future planning for space exploration in the twenty-first century.

Space has become a political horizon, providing a frontier for international cooperation. In fact, space exploration has always been a political horizon, first for international competition in the latter part of the twentieth century – now destined to be the horizon for international cooperation in the twenty-first century. Space is pristine, virgin territory – implicitly international. Many nations are now capable of participating in the space exploration enterprise and space is becoming the testing ground, a general commons, on which discourse in working and cooperating together is conducted. We compete with one another, but we also cooperate where that cooperation is in the best interest of all – very much like frontier America in the nineteenth century.

And so I believe that space exploration is about expanding horizons. And space will be the defining horizon for civilization in the twenty-first century. We are moving from a century of military confrontation and international hostilities to a century of economic competition with international cooperation. In transitioning to the New Millennium, in the twenty-first century we exit an era of exploring the last reaches of our own planet that began in the nineteenth century to an era of exploring the Solar System and universe that was begun in the twentieth century. Our duty in the last decade of the twentieth century is to ensure that we provide the basis and tools for opening a new century of space exploration when we cross the New Millennium boundary in a few years. In the last decade of the twentieth century we need to secure the foundation for the next 100 years of space exploration in the twenty-first century that our children and their children will enjoy.

In space science there are several goals that we have set for ourselves in order to lay the foundation for our enterprise in the beginning of the twenty-first century. In planetary exploration, we need to complete the reconnaissance of the Solar System with a flyby of Pluto, the one planet remaining to be explored with our robotic spacecraft. We need to begin an era of surface exploration for the most accessible of planetary bodies – asteroids, comets, and Mars in particular. We also need to establish a program to detect and study planets around the nearest stars in order to answer the question that Carl posed for us at the beginning of the Space Age: Is our solar system, and is life on planet Earth, alone in the universe? In astrophysics, we need to have in place the means to complete the initial survey of the universe across the entire electromagnetic spectrum from gamma rays to the submillimeter and begin to answer questions on the origin, evolution, and fate of the universe, galaxies, and stars. In space physics, we need to have in place the means to understand the sun–earth connection, to answer questions concerning the flow of energy from the sun to the earth's atmosphere and surface, and to understand the effect of solar variability on the earth.

In order to do all this in today's federal budget climate, we need to have a revolution in the way we do space science. We need to move away from an era in which we conduct a few, large, costly, workforce-intensive missions to an era with a much higher flight rate of smaller, cheaper, less workforce-intensive missions using new low-mass technologies to increase the per-unit-mass capability of spacecraft by orders of magnitude. We need to build spacecraft subsystems on 1-gram silicon chips instead of building 15-kilogram card cages. We need to move from the 2000-kilogram class spacecraft of the last decade, and the 250-kilogram spacecraft of today, to 10- to 50-kilogram spacecraft by the turn of the century. We need payload mass-fractions of 50% instead of 15%. Instead of building spacecraft and determining how to interface instruments with them, we need to determine what instruments are required and fly them by adding the required flight functions in a wholly integrated fashion – a concept recently given the name "sciencecraft."

This is where space science needs to be as we transition into the New Millennium. In my view, the mission of this country's space exploration enterprise, in the fewest possible words, is to explore the universe, to seek out new planets, and to search for life elsewhere in the galaxy. In fulfilling this mission, we need to use earth-orbiting spacecraft to develop the means to understand life-sustaining processes on this planet; we need to travel robotically beyond earth in our own solar neighborhood to survey, explore, and sample every accessible body in the Solar System; we need to extend our vision of planetary exploration beyond our own Solar System in order to conduct an astronomical search for planets around the nearest stars; we need to expand our ability to observe the distant universe in order to survey the universe across the entire electromagnetic spectrum; and, all the while, we need to develop the means for human exploration beyond earth orbit so that we may in the twenty-first century fulfill human destiny to explore beyond earth and to utilize the Solar System for the full benefit of the people of this planet.

2
The Search for the Origins of Life: U.S. Solar System Exploration, 1962–1994

EDWARD C. STONE

Jet Propulsion Laboratory,
California Institute of Technology

On the occasion of Carl Sagan's sixtieth trip around the Sun, I would like to describe some of the highlights of the American planetary exploration program in which he has been so deeply involved from the very beginning. That beginning was thirty-two years ago, when Mariner 2 flew by Venus; since then, NASA's automated spacecraft have sent back images and other scientific data from every planet in the Solar System, except Pluto: Mercury (the only planet that has not engaged Carl's attention in a significant way), Venus, Earth and its Moon, Mars, Jupiter, Saturn, Uranus, and Neptune (Plate I). Although Carl has a wide-ranging interest in understanding the Solar System and the cosmos, I think his deepest interest has been in understanding the origin of life, not just from the point of view of organic chemistry but from the much broader perspective of what conditions were critical factors in the origin of life and in the emergence of the human species. As a result, it seems appropriate on this occasion to focus on the broader understanding of the origin of life that has been gained from the first thirty-two years of planetary exploration.

Looking for Liquid Water in the Solar System

One of the conditions thought to be critical for the origin of life is the presence of oceans of liquid water; as Earth's twin planet, Venus seemed to be the place to look for this critical planetary feature. Venus is almost exactly the same size as our planet (Figure 2.1), with similar physical properties. But, even before the age of planetary exploration began in 1962, Carl was warning that the surface of Venus was just too warm for liquid water, as subsequent missions revealed to be the case. Even though investigations revealed a surface much too hot for liquid water, in 1970 planetary scientists continued to entertain the possibility that water ice existed in the clouds of Venus that effectively

FIGURE 2.1
Venus (left) is nearly the same size as Earth, but conditions on Earth's neighbor are not favorable for life. (All figures in the chapter with the exception of Figure 2.18 are courtesy of NASA/JPL.)

block the view of its surface. Within several years, however, even that possibility had been eliminated when data suggested that the Venusian clouds are composed of sulfuric acid.

Interest in Venusian water was reestablished later in the 1970s by the Pioneer Venus mission, which discovered an excess of deuterium, or heavy hydrogen, in the atmosphere of Venus. This discovery suggested that at one time there had been a lot of water on the planet's surface, but a runaway greenhouse effect had eventually eliminated the water through evaporation and then loss of the hydrogen at the very top of the atmosphere. This led to one of the primary objectives of the Magellan spacecraft, which recently completed its mission at Venus as it plunged into the atmosphere, to look for evidence of ancient seashores on the surface of our sister planet.

To peer through Venus' clouds, Magellan used an imaging radar system, revealing a planetary surface greatly scarred by major rift valleys and large volcanoes. Plates II and III are false-color radar images of Venus. (The colors in many of the images discussed in this article have been enhanced, and in Plate III the color is based on surface images from the Soviet Venera landers and the vertical relief has been exaggerated so that one can more clearly see the large rift zones and

FIGURE 2.2
This Viking mosaic of Mars shows the canyon system Valles Marineris (below center), which stretches more than 3,000 kilometers in length. (Image processing by U.S. Geological Survey, Flagstaff, AZ)

volcanic constructs characterizing the Venusian surface.) This surface is probably about 500 million years old; the volcanic activity has long ago obliterated evidence of ancient seashores. So, unfortunately, the planet did not tell us much about the conditions that are important to the origin of life. What Venus did teach us was that if you move a world much like Earth 28% closer to the Sun, planetary conditions become inhospitable to life.

Where else in the Solar System might one look for evidence of an ocean – at least one that existed in the past? In the 1970s, Carl, along with many other planetary scientists, turned his attention to the planet Mars (Figure 2.2). There is evidence that three billion years ago there was a great deal of water on the surface of Mars (Figure 2.3), for a long enough time for primitive life to have evolved. Unfortunately, in 1976, when two Viking spacecraft settled down at Martian sites that Carl had a lot to do with choosing (Figure 2.4), they found no evidence of organic residue on the surface; but many of us still believe that Mars

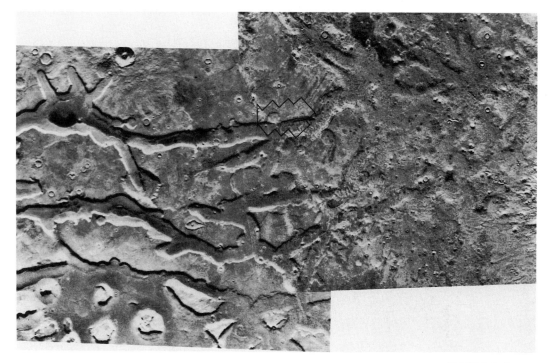

FIGURE 2.3
These channels on the surface of Mars were likely carved by ancient flows of water.

can shed a great deal of light on the conditions under which life can evolve.

Where else in the Solar System might liquid water be found? In the 1980s, we looked farther out in the Solar System, to the giant planets. The first of these is Jupiter. Figure 2.5 shows Jupiter with two of its moons: Io (left), the closest moon, which is six planetary radii from the center of Jupiter, and Europa, which is about ten planetary radii from the center of the planet. Both satellites, which are about the size of our own Moon, orbit Jupiter in synchronism, with Io orbiting twice for every one orbit of Europa. Because of the gravitational interplay of these two satellites in Jupiter's immense gravitational field, Io's crust is constantly flexed by tidal forces. This flexing generates enough heat to cause extensive volcanic activity, resulting in a constantly renewed surface that is unmarked by impact craters (Figure 2.6). The shades of orange that apparently mark Io's surface are most likely associated with the presence of sulfur, a possibility that Carl studied in some detail.

Figure 2.7 shows an interesting Io feature named Loki Patera (the crescent-shaped black spot below center), believed to be a lake of liquid sulfur with a solidified surface crust. The evidence for this is an average surface temperature on Io (which is five times farther from the Sun than Earth is) of 120 kelvins, or 120° above absolute zero, whereas the temperature of Loki Patera is 310 kelvins, warmer than room temperature. Loki Patera and the other black spots seen in the

FIGURE 2.4
Surface frost, but no evidence of organic residue, was found at this Viking Lander site.

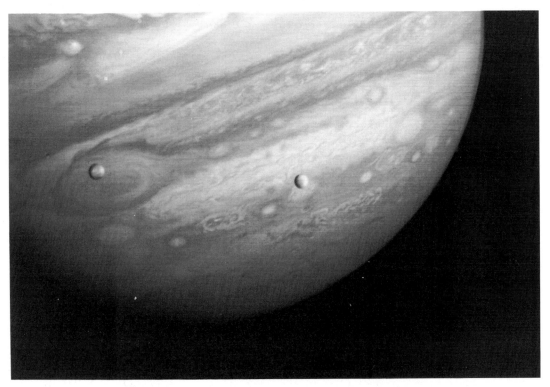

FIGURE 2.5
Jupiter, with two of its satellites: Io (left) and Europa. Io is passing over the Great Red Spot, a hurricane-like storm in the planet's atmosphere.

FIGURE 2.6
Io's surface, unmarked by impact craters, is constantly renewed by extensive volcanic activity. (Image processing by U.S. Geological Survey, Flagstaff, AZ)

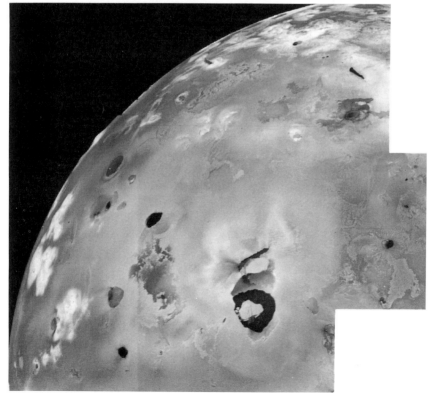

FIGURE 2.7
Io's Loki Patera, the crescent-shaped black spot below center, is thought to be a lake of liquid sulfur with a frozen surface. (Image processing by U.S. Geological Survey, Flagstaff, AZ)

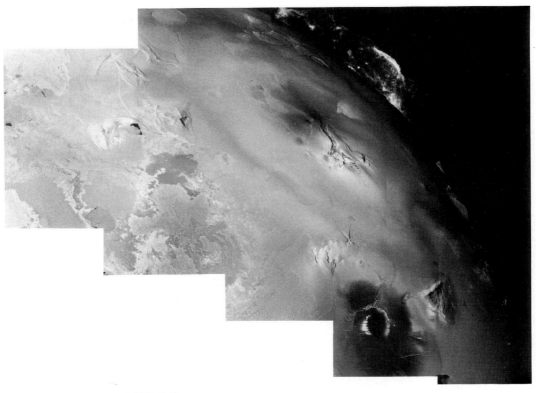

FIGURE 2.8

Visible about 300 kilometers above Io's limb is a cloud of material ejected by the volcano Pele (the large surface feature centered below the cloud). (Image processing by U.S.
Geological Survey, Flagstaff, AZ)

image are so warm that their eruptions can be observed by Earth-based
telescopes as an increase in Io's temperature, seen as a brightening in
the infrared spectrum. Although such brightenings had been observed
by astronomers before the Voyager mission, they were not understood
until Voyagers 1 and 2 flew by Jupiter in 1979.

Figure 2.8 offers a view of Io's volcanic activity. A cloud of material
ejected by the volcano Pele is visible about 300 kilometers above the
limb. (Pele is the large surface feature centered below the debris cloud.)
The plume from Pele is depositing a heart-shaped rim (Figure 2.6),
probably driven by the evaporation of liquid sulfur (rather than water
as on Earth), because there is no water left on Io. Being waterless, the
satellite reveals nothing about the conditions leading to life.

Europa is engaged in the same gravitational tug-of-war as Io, although being farther from Jupiter is less affected; as seen in Figure 2.9,
Europa has an icy crust, although it is mainly a rocky object like Io.

As the Voyager 2 spacecraft approached Europa, it revealed the
smoothest surface seen thus far in the Solar System (Figure 2.10). The
highest features on this icy surface are the narrow white streaks in
the center of the darker streaks; these white streaks are a few hundred
meters high. There are no mountains and very few impact craters,

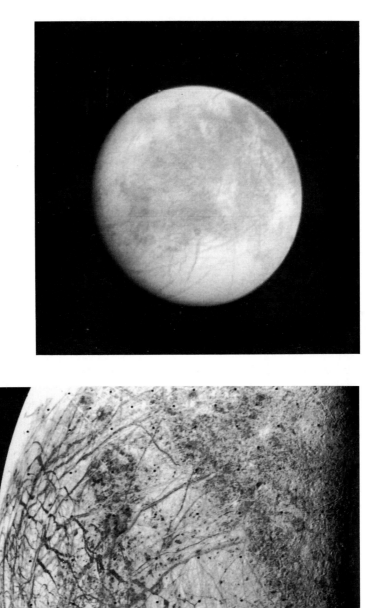

FIGURE 2.9
Unlike Io, Europa still has water – in the form of an icy crust.

FIGURE 2.10
Europa's surface is the smoothest surface observed in the Solar System: its highest features are the white streaks, which are perhaps a few hundred meters high. (Image processing by U.S. Geological Survey, Flagstaff, AZ)

indicating a very young icy surface. It is possible that beneath this surface there is an ocean of liquid water. The tidal heating that drives Io's hot spots and volcanoes may also produce enough heat to maintain an ocean of liquid water beneath Europa's icy crust. The Galileo spacecraft will fly 100 times closer to Europa than the Voyagers did in 1979 and will provide a close look at a place where there may be liquid water. If there is evidence of an ice-covered ocean, it would be exciting to send a probe to explore on and below Europa's surface.

Clues to Large-Body Collisions

Although the presence of an ocean may have been critical to the emergence of life, collisional processes can also affect the evolution of life in the Solar System. These processes were not at all understood before the Space Age began. Mimas — a small, icy world, about 400 kilometers across, in orbit around Saturn — is shown in Figure 2.11. Its heavily cratered surface is indicative of the expected effects of collisions. Collisional processes were thought of as cosmetic, producing blemishes indicative of the age of the surface, but not of fundamental importance. Figure 2.12 reveals the other side of Mimas and the large impact crater Herschel. About 130 kilometers across, Herschel has a central rebound peak some 9 kilometers high. This impact crater is so large that if the object creating it had been much larger, Mimas itself would have been fractured into many pieces. Calculations suggest that Mimas has been struck repeatedly by large bodies and broken up

FIGURE 2.11
Mimas, a satellite of Saturn, has a heavily cratered surface, indicating that the moon has been struck repeatedly by other bodies.

FIGURE 2.12
The other side of Mimas reveals the large impact crater Herschel, about 130 kilometers across.

several times. If so, the Mimas seen today is the third or fourth generation of the world that originally formed in its location in orbit around Saturn, and collisional processes are not just cosmetic, but fundamental to the physical evolution of this particular world.

Figure 2.13 shows a world even farther out in the Solar System – Miranda, a small moon (about 470 kilometers in diameter) orbiting Uranus. Another icy world, Miranda has probably the most complex surface ever observed in the Solar System, suggesting that collisional processes may well have contributed to its physical evolution.

Hyperion, another moon of Saturn, is captured in Figure 2.14. Roughly 400 kilometers across, this is a medium-sized moon that looks like the fragment of a much larger object. Figure 2.15 provides a glimpse of Saturn's rings, which also likely resulted from the collision-induced breakup of larger bodies in orbit around the planet, producing some of the beautiful complexity of structure seen in the rings. The gaps in the rings are maintained by the presence of some of the larger fragments, and the narrow band beyond the outer edge of the thicker rings seen in the photograph is contained by two shepherding moons.

It appears that collisional processes are very important to the physical evolution of bodies in the Solar System, even in the case of Earth (Figure 2.16). Planetary scientists now believe that Earth's Moon most likely was the result of the collision of a Mars-sized object with Earth – a collision that melted Earth's crust and the object itself, spinning out a ring of material that condensed to form the Moon.

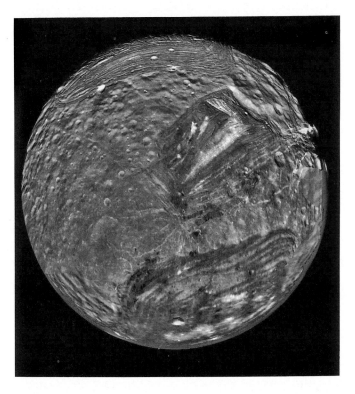

FIGURE 2.13
Uranus' moon
Miranda shows a
complex surface
that may well have
resulted from
past collisions.

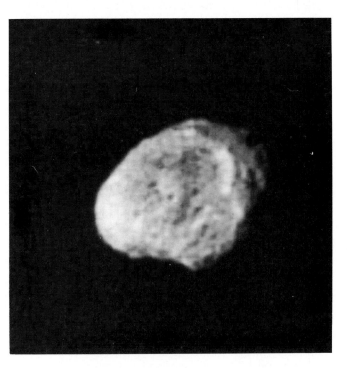

FIGURE 2.14
Hyperion, in orbit
around Saturn, is
clearly a fragment
of a much larger
object.

FIGURE 2.15
Saturn's rings are now believed to be the result of the collision-caused breakup of larger bodies in orbit around the planet.

How might this planetary collision be related to a condition leading to the origin of life on Earth? One suggestion is that the presence of the Moon and the angular momentum associated with a moon orbiting Earth have stabilized the rotation axis of our planet, thereby providing the equitable climatic conditions necessary for the evolution of complex life forms. Of course, collisions can also affect life in a negative sense. Many believe that 65 million years ago, the impact of an object with Earth led to the extinction of the dinosaurs and many other species. If true, collisional processes affected the evolution of life on Earth in a major way and perhaps created the conditions under which mammals could evolve, leading to the eventual emergence of *Homo sapiens*.

The fact that collisional processes are not just ancient effects was made dramatically clear by the impact of the many fragments of comet Shoemaker–Levy 9 with Jupiter in July 1994. Figure 2.17 is a set of images taken over seven and a half seconds by the Galileo spacecraft as it was approaching Jupiter; clearly visible to the left of Jupiter's disk is the flash of one fragment of Shoemaker–Levy 9 as it is plunging into the Jovian atmosphere. The impact site of another Shoemaker–Levy 9 fragment is seen at the bottom of the Jovian disk in Figure 2.18.

FIGURE 2.16
The presence of Earth's Moon – most likely the result of a Mars-sized object colliding with Earth – may have stabilized our planet's rotation, thereby providing the climate needed for complex life to evolve.

FIGURE 2.17
In this series of images taken by the Galileo spacecraft, a fragment of comet Shoemaker–Levy 9 – the bright spot to the left of Jupiter – is seen plunging into the planet's atmosphere.

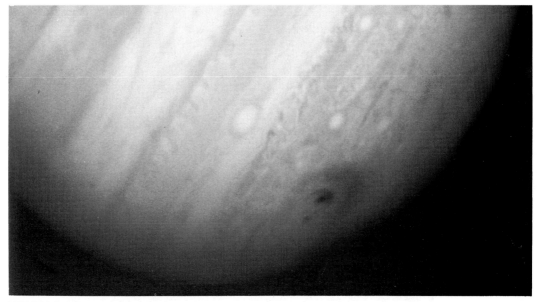

FIGURE 2.18
Jupiter displays evidence of an impact by a fragment of Shoemaker–Levy 9. (Image from
the Hubble Space Telescope's Wide Field/Planetary Camera.) (Courtesy NASA/Space Sciences Telescope Institute)

The size and orbit of Jupiter may turn out to be fortuitous in terms
of the conditions for the evolution of life on Earth. Calculations suggest
that the presence of Jupiter at 5 astronomical units from the Sun – five
times farther from the Sun than Earth – has been an important barrier
to the rain of cometary debris that would otherwise have impacted
Earth with much greater frequency, perhaps altering, in a major way,
the evolution of life on Earth.

Collisional processes that at the beginning of the Space Age were
thought to be mainly cosmetic may be instead a very important element not only in the physical evolution of the bodies in the Solar
System, but in the evolution of life itself.

Organic Building Blocks of Life

A primary consideration in trying to answer the question "What conditions led to life?" involves organic chemistry, a principal area of
Carl's research for many years. One of the bodies that caught his early
attention was the Saturn-orbiting moon Iapetus (Figure 2.19). From
its discovery, Iapetus was known to have a very peculiar property:
one side is very bright, with perhaps 50% reflectivity, and the other
side is very dark, with perhaps 5% reflectivity. Presumably, Iapetus is
composed mainly of water ice; then what is the dark black material
that covers half of its surface? If Iapetus were the only object of this

FIGURE 2.19
The black carbon-bearing material that covers half of Saturn's moon Iapetus and that appears on other bodies in the outer Solar System is still a mystery.

type in the Solar System, it could be considered a special case, but as other icy bodies in the outer Solar System were explored, planetary scientists began to realize that Iapetus is not unique.

Figure 2.20 shows Umbriel, another satellite of Uranus. This moon is grey, reflecting only about 20% of incident sunlight. In addition, the rings of Uranus are charcoal black, and we now know that the surfaces of comets such as Halley are charcoal black. What is this black material and where did it come from? Is it primordial dust material that came from the interstellar cloud out of which the Solar System formed? Or is it an organic material that has resulted from already formed bodies in the Solar System being irradiated or affected by some other processes? The answer is not yet known because we do not have a sample of this material for analysis.

However, we are beginning to obtain samples of some of the interstellar grains that may provide us with insight into this black carbon-bearing material. Observations indicate that it is spread throughout the outer Solar System on the icy bodies. We know that much of this material was deposited in Earth's atmosphere by collisional processes, raising a series of new questions: What is the role of this material in the atmosphere of early Earth? How much of it survived the entry into Earth's atmosphere to contribute to the prebiotic organic inventory in Earth's oceans? These are some of the interesting problems that will have to be answered before we can understand what are the central conditions for the origin of life.

FIGURE 2.20
Reflecting only 20% of received sunlight, the Uranian moon Umbriel is covered by the mysterious carbon-containing material.

There are some places in the Solar System where organic material is now being synthesized. Triton (Plate IV), a moon in orbit around Neptune, is the coldest body visited in the Solar System. At 30 astronomical units from the Sun, Triton has a surface temperature of 38 kelvins. It is so cold on Triton that its polar ice cap consists not of water ice, but of frozen nitrogen. The moon is about one-quarter water ice and three-quarters rock. The image reveals an icy surface with an unusual pattern. The significant thing about Triton's surface, though, is its color: not white as one would expect from a surface of water ice, but a brownish color suggestive of an organic residue. Triton has a very thin atmosphere of nitrogen and methane (the latter known as natural gas). It is likely that the methane is slowly being converted into more complex hydrocarbons that eventually polymerize and form a brownish deposit on the surface.

Let us return to Saturn. One of its moons, Titan (Figure 2.21), is an object of great interest in terms of current organic chemistry and was the focus of Carl's attention during the 1980s. A planet-sized moon, approximately the same size as Mercury, Titan differs from Mercury and some of the other planets because it has a substantial atmosphere: its atmospheric surface pressure is 1.6 times greater than that of Earth. Like Earth, Titan has an atmosphere that is 80% nitrogen; but unlike our atmosphere, it contains not oxygen, but methane.

Titan has a thick haze layer – an opaque layer of complex organic polymers whose composition is not yet known. Plate V's false-color

FIGURE 2.21
The very active organic chemistry occurring on Saturn's moon Titan may resemble some of the processes that occurred in the atmosphere of early Earth.

limb view of Titan shows several haze layers above the thicker, opaque haze layer. Each of these haze layers likely has a unique chemical composition that is not currently understood. The organic chemistry occurring on Titan today in some ways may resemble the organic chemistry that occurred in the atmosphere of early Earth before life evolved.

One of the key challenges in Voyager 1's 1980 encounter with Titan was to look for breaks in the haze layer that would allow us to peer at the surface. Carl felt this was a very important objective, although we all knew the probability of success was small. Unfortunately, nature did not cooperate in providing visible holes in Titan's haze layer, so we had to rely on other data to infer what was below this atmospheric shroud. As the scale in Figure 2.22 indicates, Titan's atmosphere is quite extended, and the thick haze layer is a little over 200 kilometers above the surface. Photochemistry suggests that one of the most abundant compounds created from Titan's methane by the action of sunlight and the energetic particles in the Saturnian magnetosphere is ethane, which at about 95 kelvins and 1.6 atmospheres at the surface of Titan would be liquid. So one might expect bodies of liquid on the surface of Titan, although Carl has pointed out that there are some serious tidal issues associated with the presence of large bodies of liquid on this moon's surface.

Also, as the polymers – that is, the particles of matter blocking our view of Titan's surface – increase in size they eventually precipitate, forming a very thick layer (500 meters or perhaps a kilometer thick)

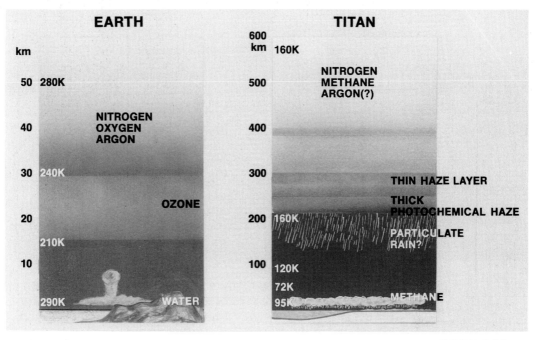

FIGURE 2.22
Evidence suggests
that a rain of
complex organic
polymers from the
thick haze layer in
Titan's atmosphere
deposits organic
residue on the
surface of the
moon.

of organic residue on the surface of the moon. This residue may be similar to Earth's polar ice caps, where the annual deposit of snowfall and the trapped gas and other constituents in each of the layers in a core sample reveal details of the past global climate of Earth. The surface residue may well hold a similar record of the organic chemistry that has occurred in Titan's atmosphere.

We will return to this fascinating world with the Cassini mission in 2004, using the Huygens probe built by the European Space Agency (ESA). The probe will drop into Titan's atmosphere and sample the products of the organic chemistry occurring there today. It is hard to imagine that Titan does not have much to teach us about the conditions and the organic chemistry that may have contributed to the origin and evolution of life on Earth. Even after the Cassini mission's visit to Titan in 2004, it seems likely that we will want to return to further explore this interesting world.

In the first thirty-two years of the age of planetary space exploration, we have indeed gained a new perspective on the circumstances, some of which may have been accidental, that might have been important to the origin of life. In his 1968 Condon Lecture, Carl pointed out that perspective is the fundamental scientific return of planetary exploration. With this in mind, I would like to close with one final example of such a new perspective – a composite portrait of the Solar System from beyond the most remote planet (Figure 2.23), created from sixty photographs taken at Carl's urging by Voyager 1 just twenty-eight years after Mariner 2's historic flyby of Venus. Of course, the

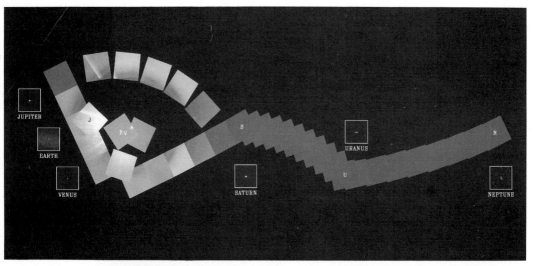

FIGURE 2.23

As Voyager 1 flew some three billion miles from Earth, the spacecraft took a series of 60 photographs that were later mosaicked into this portrait of the solar system. (Because of viewing conditions, Mercury, Mars, and Pluto were not imaged.)

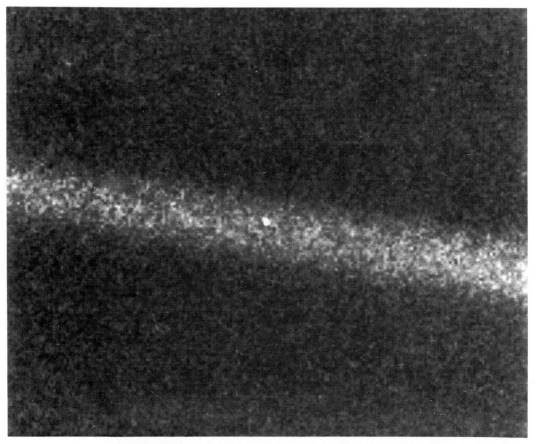

FIGURE 2.24

This enlargement of one of the Voyager 1 photographs taken from the outer threshold of the Solar System reveals Earth as a tiny blue dot (the haze of sunlight is in the camera).

Solar System is so spread out that many frames were required. In the portrait, the Sun is visible as a bright point, but the planets cannot be easily discerned, except in individual enlargements (with captions) of some of the images. Figure 2.24 shows a frame isolating Earth. In Carl's words, from this perspective, "Earth is a pale blue dot." I am confident that in the years ahead, we will continue enlarging our perspective of the origin of life as we undertake new journeys of exploration.

3

Highlights of the Russian Planetary Program

ROALD SAGDEEV

East–West Space Science Center,
University of Maryland

I have some difficulty in describing the highlights of the Russian space program. The whole transition from a program that was Soviet and now is Russian took place almost overnight, and there were no special preparations for changing the status of the program. Indeed, an interesting experiment, a live experiment was under way, on Russian, on Soviet cosmonauts, on board of Orbital Station Mir, which was launched in the prehistoric epoch, during Soviet time; then, while in orbit, they discovered that it was already the Russian epoch. In an interesting way, in an ironic way, it was probably the first experiment with living Einsteinian observers who discovered that the clocks in the orbit were moving in a different way.

So, this is why I suggest we divide the whole story into the Soviet and the Russian epochs of the space program. It's only a few years since the end of the Soviet epoch, but already it's receding back into history. As a matter of fact, when Sputnik was launched in 1957, many of us, I'm sure not only in Russia but scientists in the United States as well, thought that it would open a new era equivalent to the era of great geographic discoveries. Now, sadly enough, we Russians discovered that instead of the great geography of space, we were learning about facts of history; we made great historic discoveries of our recent past in Russia. This part of Russian history, of Soviet history, was very visible. It was a beneficiary of the Cold War. The leaders of the country very energetically promoted space exploration, I'm sure not specifically in order to help Carl and his colleagues learn more about the cosmos, but because it clearly had propagandistic importance. It was a trick to prove the superiority of the socialist system, and this is why the technique that was used was very similar to the technique of Socialist Realism. Everything about the space program was set up in the way of Socialist Realism. Announcement of a launch, about a successful launch, came only after the de facto launch took place, and small censoring eliminated those launches that were unsuccessful.

But, going back to the Soviet epoch: we had almost fifty lunar and planetary launches during a substantially short period of time, thirty-one years, essentially, beginning with the early Sputnik. Two years later, we began attempts at lunar exploration, but the first successful lunar mission took place in 1966. The reader probably remembers how everyone was excited with the first panorama of the lunar surface, taken by an imaging camera on the surface of the Luna 9 spacecraft. Before the Soviet press was able to pick up and deliver to everyone the first manmade panorama of the Moon, Bernard Lovell of the Jodrell Bank Radio Observatory picked up the telemetry data and the London *Times* published it before *Pravda*. That infuriated the Soviet leaders, of course, but they were vindicated when they discovered that Lovell did not know about the scales and in one direction his scale was stretched twofold. This cost Bernard dearly because the Soviet Academy rejected his candidacy on the basis of his interception of the lunar data.

There were twenty-four unmanned lunar spacecraft during a very short period – the last one was launched in the mid-70s. It was Luna 24, the unmanned spacecraft that was sent to return a lunar sample taken from underground from the depth of up to two meters. It was an interesting period, and now the scientific community knows much more about this particular period. At that time, even the intimate friendship I had for many years with Carl did not allow me to tell the absolute truth about the hidden side of the Soviet space program, so this may be the first time that I describe some of the details of the program, in the Socialist Pseudo-Realism version!

One particular program that is now becoming known in the West was called N1. It was a Soviet attempt to compete with the Apollo landing on the Moon. In technical terms, it was a very risky program. The big booster, the size of Saturn 5, used fifteen powerful parallel rocket boosters, and the Soviet control and electronic technology clearly could not get ready in a short time to operate and control such a complicated assembly of engines. In the period between 1967 and 1972 there were at least four major failures at launch. All these launches were test launches, unmanned, so no human lives were lost, but it was the first very serious blow to the regime, to Brezhnev, who just as Khrushchev, thought that we could easily outdo the Americans in space. It was the first blow and we Soviets understood that we were not omnipotent in space. But the feeling that every Soviet got while watching live TV pictures of Armstrong walking on the Moon was even more important. It was a feeling that we all were citizens of one planet, living on this pale blue dot. It was very important for all of us.

As a result of the failure to land on the Moon, the Soviet government had to introduce some changes in the cadre. There are no irreplaceable people, was the old motto, so the general designer of this project, academician Vasily Mishin, was removed from his position. I believe, however, that the actual explanation was not so much the failure of one program, because he was removed from his post very late, in 1974. The actual explanation was that he had a drinking problem. At that time, even in such a small position as chief designer of rockets, no one could survive having a problem like that.

In the mid seventies, the success of Viking, all the data that we were getting from Carl and his colleagues, persuaded us that we should abandon the attempt to land an unmanned module on the surface of Mars; as a consequence, we decided to abandon a lot of work that already was going on since 1961 and that resulted in several unsuccessful landings. Probably it was the first time that the scientific community was able to influence the government and persuade it to give up this area of space science. We wanted to give a chance to the scientific community to analyze the Viking data, especially in view of the dramatically decreased likelihood of finding even primitive microbial life on the surface of Mars. Instead, the government supported another idea, a sample return from Mars. Even today it sounds surreal that in 1975–6 the Soviet government, the Central Committee in the Kremlin, issued a decree to launch a Mars sample return mission. Of course, everything was done in an atmosphere of top secrecy. The mission had a coded name, 5M. No smart American clearly would be able to interpret it as a Mars sample return. The mission had a rather sophisticated technical scenario, sophisticated even for that particular epoch.

The plan was to launch two parallel Proton launchers, implement unmanned docking in orbit, in a low-Earth orbit, to build a heavy bus capable of carrying a Mars lander, and a return rocket. This bus would have to orbit around Mars and then from that orbit send a capsule, presumably the same type as the lunar capsules (we already had the technology for lunar unmanned capsules to recover lunar samples), deliver it back to the bus, and with the help of a small return rocket deliver the package back. This was an overly ambitious project. Many of us, I among them, did not believe that the state of the art in Russia, state of the art in technology, in control, would allow us to implement such a mission. What happened? Why was this mission finally abandoned about a year before we had to launch it? At that very moment, in the Soviet human flight program there were several rather noticeable setbacks. Cosmonauts launched to the orbital station Salyut, a predecessor to Mir, were unable to dock with the station. Especially dramatic was the case when the best brother of Soviet Cosmonaut Bulgarin was launched and we were unable to deliver him to the Salyut space station. It created such a shock through the whole space hierarchy that the project of complicated unmanned docking of two Protons in orbit was finally rejected – nostalgic memory of this particular epoch.

The next episode in the surrealist program of Russia was a project known to many in the United States. We were planning to launch a huge atmospheric balloon to Venus. The date of launch was chosen – 1983. We were doing it together with the French space community, and the idea was that we would celebrate the 200-year anniversary of the Montgolfier brothers' balloon flight in the atmosphere of the Earth. So, this spirit of celebration, of launching at the important historic date, prevailed, but what happened? The project technically was feasible. I'm sure we would still be getting a lot of data from this flight, which was to carry a rather sophisticated scientific payload. Part of it was to be used to study the complicated chemistry of the clouds in the atmosphere of Venus. The reason this project was abandoned almost at the last minute in 1981 was that we discovered that Bruce Murray was lagging behind schedule in trying to persuade NASA and the administration in this country to launch a spacecraft to encounter Halley's Comet. Such factors influenced our program so much that we decided to abandon the big balloon celebrating the Montgolfier brothers, but rather to send a spacecraft first to Venus with small balloons and a lander and then after a swing by to encounter Halley's Comet. I think it was probably the biggest international project, involving about nine nations, in the history of the Soviet program. At that time, not even NASA was involved in such extensive international cooperation, and it brings me to my very first memory of meeting Carl.

I heard, of course, a lot about Carl. I read some of the books. One of the books he coauthored with a very close friend of mine and a leading scientist of the Space Research Institute, Iosef Shklovskii. And then in 1976 I was planning to go to the United States and hoping to see Carl during my visit at Cornell, but he was absent from his office at that particular period, so finally we were able to arrange the meeting at the National Airport in Washington. To identify Carl, I was given a small photograph. One probably would not be able to find a single living creature on the planet now who would need a photograph to recognize Carl. It was before the *Cosmos* series. So I kept the photograph and he was very easy to find. What happened then seemed like science fiction. I met the man for the first time, and half an hour later we were talking about how to bring real openness in the Soviet program. We elaborated on a scenario in which one of the next Soviet Veneras landing on the surface of Venus would be coordinated in terms of scientific coverage of the mission, with a parallel American Pioneer Venus mission. Unfortunately, we lived in a different epoch, and only part of this scenario materialized, but a very few years later, in 1986, we fully recovered this scenario when Carl came to Moscow during a real-time encounter with Halley's Comet. I think it was such an unusual coincidence. We did not try to respect any important Bolshevik dates. After all, we had the excuse that the orbits of comets are controlled by God. But at the very moment when Carl was in real time sitting in the control room and talking to Ted Koppell, *Nightline* showed here in the United States all the pictures of the comet. At the very same time, Mikhail Gorbachev was giving the concluding remarks at the Twenty-seventh Party Congress. Two weeks later, he was puzzled. He said, "How did you manage to do it?" I said, "We had a perfect alibi. God controls orbits." He thought for a second, and said, "That means that God is with us." How ironic it was that in 1991 Yeltsin in the presence of Gorbachev after the failed coup, signed the decree prohibiting the Communist Party of the Soviet Union.

The next step in this program was the invention of Mikhail Gorbachev personally. It was a project to begin preparation for a human expedition to Mars as early as 1988. In fact, we had the technology. Two Energia boosters coupled together could do the whole job. The feasibility study was done by the Russian space industry, two Energia launchers were built, a huge heavy bus to be placed in low-Earth orbit and to be sent off by nuclear propulsion. From that point we would be able to deliver the bus with the expedition consisting of an international crew to the orbit around Mars. Again, with the Apollo scenario revisited, going to the surface and then back. And there would be no better historic circumstance, a better chance for Gorbachev to try to involve the United States in this spectacular project than walking Ronald Reagan through the Kremlin yard, showing important historic marks of Russian achievements in the past. Indeed, he brought President Reagan to see the Czar Cannon, to prove that the Russian genius was alive even in the sixteenth century, and he said, "Why don't we send a mission to Mars?" Now, I have only one complaint about Carl, and I think even the celebration of his birthday should not prevent me from being critical. Here is what happened next. The same evening after visiting the Czar Cannon and the Czar Bell, Gorbachev was giving a big party in the Kremlin, a state dinner. I was standing in the line, and then I approached Gorbachev. He seized my hand and, introducing me to President Reagan, said, "This is the man who is promoting this mission to Mars." And then, after a while, he said, "You know who in America is doing the same, who is his closest friend?" At this very moment, you know, I saw some kind of sparks in the eyes of President Reagan. I thought, great, now we will work this out at this very moment. But instead of waiting for me to suggest that it was General Abrahamson, Gorbachev said, "Carl Sagan." It was the end of this project.

Unfortunately, the Russian space program, the planetary program, is not yet born. Mars 1992 was postponed to Mars 1994, and now perhaps there would be an attempt to launch it in 1996. I hope very much that this program will have a rebirth, by maybe a miracle, like the miracle that saved the ruble overnight, from 4,000 to 3,000. So, what kind of speculations can I present? There are a lot of people working behind the scenes to materialize at least some of the following projects. At last, we have very strong joint cooperative efforts, chaired by Vice President Gore on the American side and Prime Minister Chernomyrdin on the Russian side. One particular chance is to have Mars 1998. I'm sure Bruce Murray will discuss this in conjunction with NASA's plans to launch unmanned spacecraft. There are a number of proposals. It is the first time that a group of scientists could send their proposals in open competition, in open peace. Out of about a hundred original Discovery-class proposals, at least half a dozen of the proposed missions are associated with Russian participation. Maybe in the future some of them would have a chance.

Among them there is a Venus Revisited joint capsule delivered to the surface of Venus. Then there is a lunar orbiter using debris of the Strategic Defense Initiative (SDI) exotic technology, a proton beam, to activate the lunar soil and then to pick up gamma quanta and to build a geochemical map of the moon using gamma spectrometry, active gamma spectrometry. There is a lunar landing. I discovered recently that one of the university groups was suggesting an unmanned landing with Russian cooperation. There is a Phobos sample return, using part of the Russian technology that was developed for Mars 5M and for two missions that were failures. It was a very sad episode in my own life in space research. The mystery of the loss of the second Phobos is still unsolved. At the time that it happened, we still had the secrecy of the old regime in spite of every effort. The last initiative was launched by the Los Angeles *Unsolved Mysteries* program. I had phone calls from them; they wanted to pursue this story, and the only disagreement that finally led to the collapse of this project was that they thought that the explanation would be in terms of these strange mysterious figures, structures, on the surface of Mars. I said that I would promote a different explanation, so they lost interest.

Then Fire and Ice is close to becoming a very realistic project, a couple of launches. Then Mars together, and then just recently I have learned about a new proposal, and I have to warn Carl seriously about it. We all know him as a champion of international cooperation, and he succeeded tremendously, with the present openness, with bringing Russians to be a part of the international program, but now there is a serious competitor. There was a meeting in Chelyabinsk 70. This is a Russian counterpart to Livermore National Laboratory finished in the fall of 1994. Fortunately, The Planetary Society also was represented. At this meeting, Dr. Taylor suggested a joint mission with the Russian nuclear establishment to develop nuclear warhead technology to deflect asteroids. As sad as it sounds, he might become Carl Sagan's competitor. SAD is my acronym for Strategic Asteroid Defense.

The collapse of the Soviet Union and the near-death of Soviet planetary exploration is not a purely national issue of the former Soviet Union or Russia. It touches the rest of us, the whole scientific space community and the world in a very ironic way. The space program benefited greatly from the historic space race during the Cold War. If we cannot bring another paradigm of international cooperation for the space program, for the planetary program of this country, which produced such wonderful results, miracles described in the paper of Ed Stone, the paradigm that would take over would be called "The Loneliness of the Long-Distance Runner."

4
From the Eyepiece to the Footpad: The Search for Life on Mars

BRUCE MURRAY

Division of Geological &
Planetary Sciences
California Institute of Technology

I would like to amplify Roald Sagdeev's remarks about how the Soviet space program was so driven by milestones and nationalistic symbolism. Sagdeev was diplomatic enough not to refer to the U.S. program in the same way. However, the target date for the Viking Lander to go down to the surface of Mars was July 4th, 1976, our 200th anniversary. The landing actually was on the 20th of July, for good technical reasons. NASA had the wisdom to back off from the symbolic date to be sure we were technically ready. Thus, nationalistic fervor over space activities was strong also in the United States, but being a more open and pluralistic society, scientists and engineers had a greater influence on policy and technical decisions.

U.S. space efforts have been transformed by the end of the Cold War also. It's over, really over. We're all together in a new era of rapidly evolving international cooperation and competition. Now we have to create a new, more international paradigm to motivate ourselves for real space achievement once again. That is what Carl and I and The Planetary Society have tried to focus on. How do we go beyond the bittersweet highs and lows of the Cold War into a new era that will also be distinguished by great scientific accomplishments? How can our children and grandchildren leave their mark on history that will be remembered long into the future? I hope that this conference helps to bring that need and opportunity into focus.

Introduction

My task is to talk briefly about a big subject – Mars. Fortunately, there is a powerful integrating theme for Mars. It is the search for life. "From the Eyepiece to the Footpad" refers to the allure of life on Mars from the beginning of observing Mars through telescopes in the nineteenth century to the motivation for humans to arrive there in the twenty-first century. And the single most important person for the search for life on Mars has been Carl Sagan. So it is an especially appropriate theme today.

Furthermore, the issue of life on Mars has tied Sagan and Murray together intellectually for thirty-four years – like an odd pair of siblings joined at their navels. We came from very different perspectives; I'm a geologist. I look upon life as originating in a geological context. Carl came from the life-oriented view. He looked at life as the central fact – everything else as context. It took us about twenty years to sort those differing perspectives out. We had many, many personal interactions, sometimes heated. Carl is extremely poised. He can take abuse incredibly well. But at one point in our journey toward consensus, I must have gotten to him. He snapped, "You at Caltech live on the side of pessimism." What he meant was that I focused on the observations and the limitations they impose on the possibilities. It's the Napoleonic Code of science. Facts are wrong until proven right. And I thought, "You at Cornell live on the side of optimism." Push an idea as far as it might go. Stretch the limits of the imagination. Theories are right until proven wrong. The English Common Law tradition. I think we were both right. I think that dualism illustrates a basic property of science. I might add that as in any good novel, the characters evolve, and they converge to a shared reality. That's exactly what happened to us. One of the products of that creative tension and evolution has been The Planetary Society, as well as a very rich personal friendship.

An Earth-like Mars

Nineteenth century astronomers synthesized their eyepiece observations in shaded drawings. They recorded light and dark markings on Mars. Some recorded occasional linear features. But Percival Lowell, the most influential to the public, became convinced of the existence of an extraordinary network of narrow dark lines, the sharpness and clarity of which, for him, increased with each new map. He became obsessed. He hypothesized that this network was the relict of a great system of canals built by an extinct intelligent civilization that had existed on Mars until a gradual planetary drying destroyed it. That was very, very powerful stuff for the popular mind. It led to H. G. Wells' *The War of the Worlds* and to Ray Bradbury's *The Martian Chronicles*, tremendously important cultural and literary creations.

Few scientists accepted Lowell's inhabited Mars – there were many common sense arguments against it. But the idea of plant life there remained a plausible expectation from the telescopic view of Mars.

Plate VI shows Hubble Space Telescope images of Mars. What you see in this figure is what people like Lowell were seeing through telescopes on occasions of very good atmospheric conditions. The planet in this image has a conspicuous south polar cap. That white stuff changes in proportion to the seasons on Mars just like the winter and summer cycles on the Earth. The cap shrinks to a small residual part by the end of summer in that hemisphere. In the meantime, a complementary cap develops in the north. And so the cycle proceeds: winter, spring, summer, and fall. The caps migrate back and forth, somewhat like Earth would appear if observed from Mars. Finally, the dark markings also evident in Plate VI actually undergo contrast changes and sometimes changes in shape, also on a seasonal basis.

Mars is the only planet at all like Earth in its seasons. Indeed, the tilt of Mars' spin axis relative to its orbit around the Sun – which is what causes the seasons – is within half a degree of the Earth's. That is an extraordinary similarity. No other planet is close to Earth's obliquity. In addition, Mars' rate of rotation on its axis is just like Earth. If you time the length of time it takes for a surface marking to rotate around, it's within thirty-five minutes the length of the day of the Earth. Again, no other planet is anywhere close.

So, it was a very appealing idea that Mars is the twin of the Earth, with water frost caps on Mars moving back and forth. Even in Lowell's time, it was obvious that Mars had a rarefied atmosphere. Some estimated it at 10% of the Earth's atmosphere. But there wasn't a good way of measuring just how rarefied it really was. So it was plausible that there is moisture in the soil and that the changes in contrast between light and dark areas are due to blooming of vegetation in the springtime in the appropriate hemisphere.

Additional quantitative support for seasonal plant activity on Mars was supplied in 1956 and 1960 by the leading astronomical infrared spectroscopist of the time using the 200-inch telescope. He put the instrument aperture on the dark areas and made spectral measurements and then repeated the process observing the lighter areas. He then compared their spectra carefully and found a slight but very exciting difference. The dark areas exhibited faint absorption bands around 3.5 microns wavelength in the invisible infrared. They corresponded, as well as he could tell then, to the spectra associated with chlorophyll in plants! So he really had it – positive evidence of plant life on Mars in the dark areas, just as expected. This was the most plausible interpretation in 1960, when Carl and I first became associates. It was also when space exploration was just getting started. The search for life on Mars naturally became the central theme of U.S. planetary exploration.

I better point out now how wrong scientists were about Mars in 1960. First, the white stuff is not water ice; it's dry ice, solid carbon dioxide! It's so cold and so dry on Mars that no liquid water exists at all on the surface and it hasn't for billions of years! The white stuff going back and forth is carbon dioxide freezing out seasonally of the thin atmosphere, which is itself more than 90% carbon dioxide. There is no oxygen gas in the Mars atmosphere and only a small amount of nitrogen. The total pressure is not 10% of Earth's – it is equivalent to less than 0.1%! That is the pressure at an altitude of about 130,000 feet above the Earth's surface.

The seasonal changes in dark markings are due mainly to dust and atmospheric effects – they are not vegetative activity. (Carl was one of the first people to propose that nonbiological explanation.)

The orientation of Mars' spin axis that controls these seasonal changes, which is so close to Earth's now – $24\frac{1}{2}^\circ$ – in fact, varies from up to at least 40°, and way back down to perhaps 15° or less over periods of hundreds of thousands to millions of years. It's a cosmic coincidence that we happen to be looking at it now when it is very close to our own. If *Homo sapiens* had emerged from Africa a few hundred thousand years earlier, we would not have made that mistake. If civilization had developed a similar amount later in the future, we likewise would not have been led to judge the two planets as so similar.

And there is more. The rate of rotation (length of day) is yet another coincidence! The Earth's rate of rotation has been changing due to the gravitational interaction with our large Moon, while Mars' rotation has been nearly constant. Finally, the infrared bands are not due to chlorophyll or organic matter or to anything on Mars. They are caused by water vapor in Earth's atmosphere. It turns out the spectra were taken from the dark areas and the light areas on *different* days. The ratio between them exacerbated very slight daily differences in water content in the Earth's atmosphere. Even though he was the best infrared astronomer in the world at that time, using the best telescope, still he got the wrong answer. So, all of that information on which this enormous enthusiasm for life on Mars was built up was flawed.

A Moon-like Mars

The expectation of an Earth-like Mars was still very high in July 1965 when Mariner 4 made the first flyby of Mars. As a junior member of the imaging team, I had been studying how to recognize with our very primitive television camera significant features like folded sedimentary layers remaining from hypothetical ancient oceans on Mars. Instead, we found giant craters – like the Moon. Figure 4.1 is the best of the twenty-one tiny framelets broadcast back by Mariner 4. You can see the edge of a huge impact crater. It's about 300 kilometers across. The significance of that large a crater dominating our first tiny glimpse of Mars's surface up close is profound. We knew from our general solar system knowledge that the giant impacts necessary to produce a crater that size had only happened billions of years ago, like on the Moon, not more recently. So our first glimpse at Mars is of a very ancient surface, like the lunar highlands. Equally significant, that kind of old topography gets scraped off the Earth in a hundred million years or less because of our aqueous atmosphere. So we found a *fossil* surface

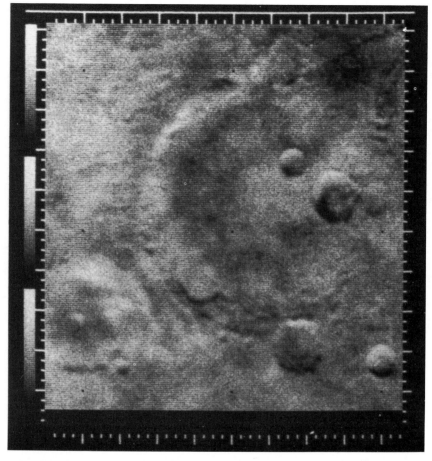

FIGURE 4.1

Mariner 4's first glimpse of Mars' surface. The large crater that dominates this tiny snapshot of Mars' surface was an astonishing clue to the fact that Mars' surface was more like the Moon than the Earth. In this frame, the sunlight is coming from the bottom of the picture. In this primitive digital television picture, there are only 200 picture elements in each direction.

on Mars, which meant that there had been no Earth-like erosion and weathering for billions of years and, therefore, no oceans, rainfall, and rivers. We knew right then, from this primitive set of pictures, that Mars was not like the Earth. It didn't have an Earth-like history. It seemed more like the Moon with a thin atmosphere at this point.

Naturally, the expectation of life on Mars plummeted. And there was more to come. My Caltech colleague, Professor Robert Leighton, puzzled about this very thin carbon dioxide atmosphere surrounding Mars. He made some basic energy calculations and demonstrated clearly that the physical consequence of that carbon dioxide atmosphere on Mars is that the frost cap should be carbon dioxide, not water ice! Then, Mariners 6 and 7 flew by Mars in 1969 and reaffirmed

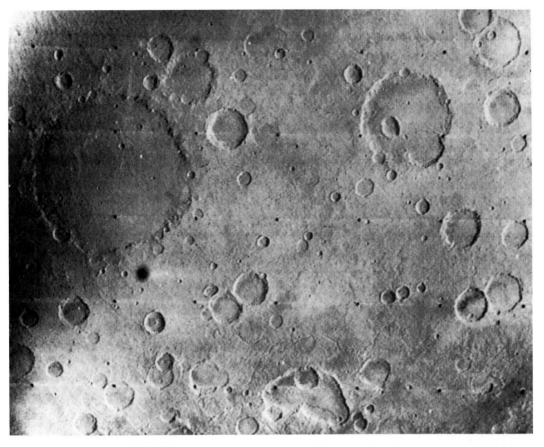

FIGURE 4.2

Mariner 6's better view of Mars' cratered surface. This greatly improved image of the surface of Mars, acquired by Mariner 6 in 1969, reinforced the view of a lunar-like Martian surface. The sun is coming from the right-hand side in this image. However, in the lower center area, there are some faint gullylike features. The significance of these was not evident until the later Mariner 9 and especially Viking images illustrated a network of these, which is testimony to an early episode of ground water erosion on Mars.

that cratered surface (Figure 4.2). However, the better camera systems showed that the craters had been blanketed and smoothed. Now we could see the handiwork of an *ancient* thick atmosphere, not the thin one that's there now, but an ancient one. Most important, Mariner 7 had been targeted to fly near the edge of the retreating south polar cap (Figure 4.3). The spacecraft carried an infrared radiometer developed by Dr. Gerry Neugebauer, also from Caltech, as well as an infrared spectrometer developed by a group at the University of California. Indeed, very, very dry carbon dioxide ice comprising the seasonal cap was confirmed by these measurements. So the nail was really put in the coffin of life on Mars in 1969. Many who had been hopeful of life were pretty discouraged, because a frozen, dry moon with a carbon dioxide envelope seemed to be a very unpromising abode for life.

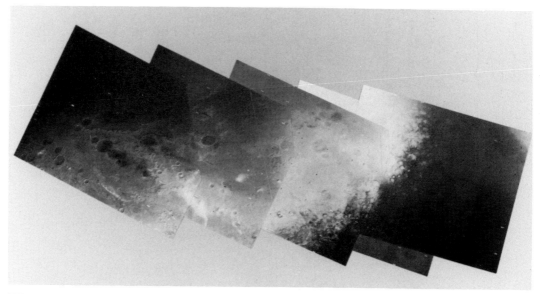

FIGURE 4.3

Mariner 7's view of the carbon dioxide frost cap surrounding Mars' South Pole. This mosaic of frames acquired by Mariner 7 as it flew over the receding edge of the Martian south polar cap in August 1969 provided strong visual evidence of the thin layer of seasonal frost. The same spacecraft carried an infrared spectrometer and photometer that proved conclusively that the white substance was frozen carbon dioxide, not water ice as many had previously believed.

A Watery Mars Past Revealed

Like most stories, the next chapter brings more surprises. Systematic observations from orbit were carried out by Mariner 9 in 1971–2. Carl and I were both members of the imaging team. Unfortunately, Mars was in the midst of a giant dust storm when we got there, as shown in Figure 4.4, a highly processed picture using the best technology at the time. It is actually four frames put together. All you could see were these funny dark spots, four of them – one in the upper left, three others lined up on the right. What in the world could they be? They proved to be huge volcanoes! That upper left-hand spot proved to be the largest volcano in the solar system (Plate VII)! It is about 500 kilometers across, large enough to occupy most of the northeastern United States. It's called Olympus Mons. So this planet is not the Moon, this has bigger volcanoes by far than the Earth. It is an active planet.

There was more to come. Up in the polar regions, we discovered very, very thin uniform layers (Figure 4.5). They are nearly flat-lying, so they tend to show up as curvilinear surface patterns of low relief. These enormous layers pretty clearly chronicled global climate changes. There are the same kinds of layered features right among the south polar ice. So Mars has had enormous climatic fluctuations quite

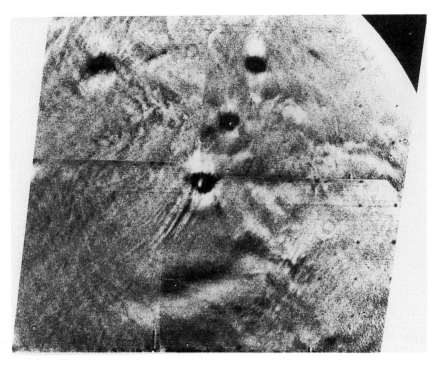

FIGURE 4.4

Through a dust storm darkly. Mariner 9 arrived at Mars at the time of a great dust storm and, even with the best available processing at that time, this mosaic of four pictures reveals little more than the limb (upper right) and four dark spots. As the dust cleared, it became evident that the four dark spots were the summits of huge volcanoes, larger than any similar feature on the Earth, that were so high that they protruded through most of the dust storm. The dark spot on the upper left is Olympus Mons, which is shown in clear form in Plate VII.

unlike the Moon! Something else has been going on, perhaps analogous to glacial epochs on Earth. Perhaps in those climatic fluctuations, conditions got much more amiable and life could survive.

But the biggest surprise were the channels, not the imaginary ones that Lowell was talking about, but great huge gouges in the surface of the planet created in ancient times by catastrophic flooding (Figure 4.6). Furthermore, the enormous craters we first recognized with Mariner 4, and then saw again with Mariner 6, better cameras now showed to have many small gullies carved in their surface. Apparently there was ground water erosion shortly after they formed. The great channels likewise were carved billions of years ago – after the huge craters, but before the emplacement of huge plains of lava.

The channels completely changed the game. Mars' surface records an ancient, but very powerful, aqueous history. So the idea that life once formed there became plausible. Maybe some microbial forms even survived to the hostile present. Carl was a leader in that line of reasoning.

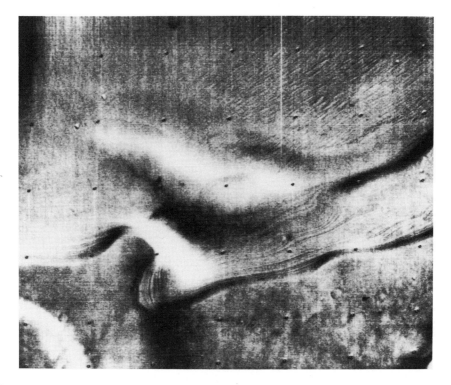

FIGURE 4.5

South polar layered terrains. This preliminary image from Mariner 7 shows the eroding
edge of a delicately layered blanket of wind-deposited sediments that characterize both
the north and south polar regions of Mars. The sun is coming from the lower left in this
image. The regular pattern of black dots is geometric reference points in the focal plate.
At the lower portion of the picture, faint craters can be seen from the underlying surface,
beneath the blanket. The thin layers are part of an escarpment that is facing equatorward
(north). The upper portion of the picture is dominated by a lower left/upper right set of
streaks that are believed to be a wind erosional feature called yardangs, carved out in a
very smooth top surface of this blanket of wind-deposited material. This image is about
50 kilometers (30 miles) in horizontal dimension.

The Viking Search For Life

Timed to coincide with the U.S. bicentennial in 1976, the most elaborate unmanned mission ever deployed arrived at Mars. Figure 4.7
shows a model of the Viking lander. Note the long arm designed to
reach out and sample the soil, then drop the samples into a special
processing laboratory aboard the sophisticated lander. Included in that
laboratory were some of the most sophisticated life-detecting experiments ever built.

How does one detect life? It's a tough question. If somebody gives
you some stuff, how can you prove something in it is alive? There was
only one really universal way. One has to find evidence that something
is replicating and growing. There were three experiments to look for
something growing. Two of them used differing liquid broths, offering

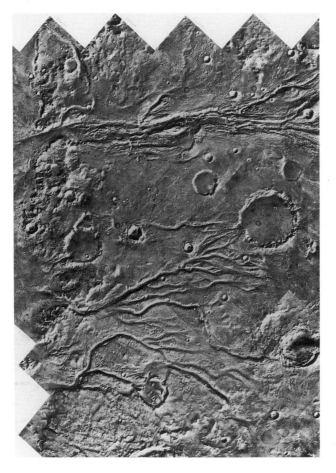

FIGURE 4.6

A rich aqueous history on Mars is chronicled by this mosaic of Viking images centered at around 17°N, 55°W. This is west of the Viking 1 landing site (to the left of this image). This terrain in the picture slopes from west to east with a drop of about 3 kilometers. The channels are a continuation of those to the west and formed perhaps three billion years ago when catastrophic flooding from subterranean sources carved the equatorial regions of Mars extensively. The sun is coming from the lower right in this mosaic.

something that putative Martian bugs might like to eat. Then chemical changes resulting from metabolic activity were measured to prove growth – if any. The third experiment measured the uptake of radioactively tagged carbon dioxide in a nearly dry experiment.

Furthermore, if there is life, there obviously must be associated organic material in the soil in which microbes exist. Perhaps the bugs themselves, but certainly their carcasses and residual chemical environment, must be there at least in trace amounts. So there was another experiment called a mass spectrometer/gas chromatograph (GCMS), which was an extremely sensitive way to look for the organic material in the soil.

FIGURE 4.7

A drawing of the Viking Lander as it was imagined to look on Mars after having landed. The entire structure is about a meter and a half tall and in excess of 2 meters in horizontal dimension. The important element is the arm extending from the right-hand side of this model. The lander contained a sophisticated biological laboratory operated remotely, which was capable of detecting any simple life-forms. It also contained a very powerful analytical instrument that was capable of finding traces of organic materials down to one part in a billion.

Finally, the day came when the arm went out, digging and scraping up a soil sample (Figure 4.8). That soil was put into the Lander and the first test using the broth was carried out. There was an extraordinary result. The data ran off scale! There was a tremendous evolution of gases. It was quickly realized that this was not a biological effect but an inorganic chemical one. There was some natural compound in that Martian soil that was actually breaking down Viking's liquid reagents.

This "unearthly" superoxidizing process was confirmed by the GCMS. When it analyzed for organic materials – from simple hydrocarbons to more complex compounds – it found none at all! The GCMS had extremely high sensitivities. So we realized that the surface of Mars as it presently is, is self-sterilizing. Any traces of organic material, even that brought by meteorites and comets, doesn't last long on Mars. Otherwise we would have seen it.

In fact, Mars has a superoxidizing soil, a situation that can't exist on the Earth because we have water vapor all around. Any microbes on

FIGURE 4.8

The Viking Lander collects a soil sample to test for microbial life. The left-hand view shows the arm deployed and trenching in behind a rock on the Martian surface to collect a sample. The right-hand side shows the resulting trench from which the sample was collected. No life was found in any such samples by either Viking 1 or Viking 2. The sophisticated analytical equipment discovered that the surface was devoid of organic material to an unprecedented level. Thus, the surface of Mars is sterile. It is believed that this circumstance results from the fact that hard ultraviolet radiation reaches the surface of Mars directly, unlike the Earth, which protects us from this consequence with the ozone layer.

Mars would be chemically destroyed just the way Viking's broth was. In the case of Mars, very strong solar ultraviolet radiation reaches the surface. On Earth, our ozone layer protects us. The strong ultraviolet radiation breaks up the water vapor molecules and also acts on the mineral grains. Altogether it makes the environment quite reactive – a superoxidizing situation. An analogy on Earth would be if the antiseptic hydrogen peroxide were naturally produced in our ordinary environment. Our hair would be slowly dyed automatically. Our clothes would be gradually bleached. Our environment would become automatically sterilized.

What's Next?

So once again the prospect of life now on Mars' surface plummeted. Yet Mars may have supported at least simple life forms in the geological past. This is where the convergence of Sagan and Murray came about.

Carl's original optimism about finding a living biota on Mars that we could sample and compare with Earth's faded. But despite my original pessimism, there was overwhelming evidence that Mars once had a very active early history with a lot of water present on the surface in different forms. That time scale seems long enough for life to have formed on Mars analogously to how it must have on the Earth. The implication is that when Mars changed its atmospheric state, probably about three billion years ago, to an extremely hostile one, any surface life that might have developed was not able to survive.

So the post-Viking scientific focus has become the search for tell-tale clues to past life. On Earth, the most abundant clues to past life are indirect ones – geological strata high in biogenically deposited calcium carbonate, and best, those that contain organic matter. If similar layers of ancient calcium carbonate were discovered on Mars, it would be suggestive of an earlier life-bearing epoch. But, the definitive indirect proof would be discovery of deposits of ancient organic matter. Inasmuch as the present surface environment of Mars destroys organic matter as revealed by the Viking lander experiments, we need to know how deep do we have to go to get below this oxidizing layer. A meter? Ten meters? A hundred meters? That is a key element. Surely there must be on Mars somewhere relatively uncontaminated, undestroyed strata surviving from that early aqueous period. Such locations are what we want to find, especially near enough to the surface that advanced robotic drilling or other systems could search directly for organic material. The full exploitation of such sites could become the scientific objective of future human missions.

Building for the Future

Following Viking, and the extraordinary exploration of the outer Solar System with the two Voyager spacecraft, the U.S. planetary program collapsed. NASA had become completely committed to a technological objective, the Space Shuttle, rather than a scientific one. NASA became focused on the means rather than the ends. Space shuttle development overcame the agency. It was a bad choice to put a human in the loop for everything in space. The United States has abandoned that approach after Challenger, but it cost the U.S. space program nearly fifteen years of progress. The most critical damage was to robotic deep space missions that required the greatest propulsion to get to the planets.

Carl was back in Pasadena for Voyager in 1979–80. He's a serious participant in planetary exploration. He isn't a weekend warrior. He moved out there. We spent considerable time together at that point. Both of us had became convinced from our independent and somewhat overlapping experiences, that the enthusiasm, commitment, and hope of people generally for planetary exploration, for the search for extraterrestrial intelligence, remained very, very strong. Yet it wasn't being manifested in the NASA that existed then.

So, Carl and I organized The Planetary Society, joined quickly by Dr. Louis Friedman, the third member of our triumvirate.

The search for life on Mars brought us together as close friends and collaborators in exploration. The Planetary Society was the outcome. We invented a nongovernmental entity that could precede governments in developing space cooperation. That was why Carl and I and Lou spent a lot of time, especially during the Soviet era and in the transition to the Russian era, out in front of the U.S. government. Simultaneously, Roald Sagdeev pushed international cooperation from a personally more dangerous position. He had to do it within the Soviet restrictions that existed. But we had the chance to operate as a U.S. nonprofit organization, and we did. The Planetary Society organized and sustained a whole series of collaborative Mars exploration projects. These efforts included French, Soviet, and U.S. development of a novel Mars Balloon and instrumented guiderope. Major efforts were devoted to international field testing and development of a prototype Russian Mars Rover. Furthermore, in 1985, The Planetary Society began U.S. and international advocacy of the goal of international human flight to Mars during the first quarter of the next century. That goal remains the unifying long-range goal of current Mars exploration efforts. This nongovernmental organization, with a hundred thousand dedicated members spending their own money, has made a difference in many ways, but especially in the case of Mars. The fact that there is a tangible future on Mars now – after all the setbacks and disappointments – is partly due to The Planetary Society.

Where does all this lead? The robotic exploration of Mars is center stage on the international space community's agenda. The United States and Russia are readying missions for launch in 1996. The Japanese and the United States are developing missions that will be launched in late 1998, and early 1999. The long-term goal of international Mars human exploration, I think, is a widely shared one.

The search for life on Mars started out at the eyepiece, full of misconceptions, which were revealed by early space flights. New misconceptions cropped up, which were, in turn, clarified. The progenitor of the theme of life on Mars has been Carl Sagan. The unifying theme of this symposium is the search for life elsewhere, and the enhancement of life here. That captures the essence of Carl Sagan.

LIFE IN THE COSMOS

5
Environments of Earth and Other Worlds

OWEN B. TOON

Earth Science Division,
NASA Ames Research Center
and Laboratory for Atmospheric
and Space Physics
University of Colorado, Boulder

Before the spacecraft era the planets were veiled in mystery, glimpsed just dimly by astronomers through the telescope. Even for Mars, the closest planet whose surface could be seen, the best Earth-based images revealed only large-scale markings such as polar caps. These early observations raised many intriguing questions. Were there canals on Mars? Or perhaps the long thin markings noted by some astronomers during short periods of good seeing, but never photographed, were just wishful thinking? Could the seasonal variations in the bright and dark markings on the surface of Mars be the spring bloom and fall decay of vegetation? Or were the changes in these markings due to an atmospheric phenomenon – just the growth and decline of gigantic dust storms on a dry and lifeless planet? Might the moons of Mars be hollow shells – perhaps orbiting spacecraft sent by a distant and unknown civilization? Or were the moons little chunks of rock whose density had been measured incorrectly?

Whereas astronomers obtained increasingly sophisticated data to resolve these issues, to the average person the planets remained mere figments of the imagination. The planets seemed no more real than Atlantis or the worlds constructed by science fiction writers such as Jules Verne or Edgar Rice Burroughs.

A map representing Edgar Rice Burroughs' vision of Mars has hung on the hallway wall outside Carl Sagan's office for more than twenty years. Why would this science fiction view of Mars be of such enduring fascination to a scientist? Of course, astronomers have vast imaginations. Mars may be lifeless, but there are countless planets yet to be discovered and it is fun to think of encountering Dejah Thoris, the voluptuous egg-laying Martian princess, or the exotic monsters who are depicted around the periphery of the map of Mars. Then too the map is a tribute to the ingenuity and foresight of earlier generations. The map includes an atmosphere-generating station, correctly recognizing the tenuous nature of the atmosphere of Mars and the possible need to terraform the planet to maintain life as we know it. A giant mountain fictionally mapped near the Martian equator presages the discovery of Mt. Olympus, the largest known volcano in the solar system. After Mariner 9 photographically mapped Mars, Carl Sagan showed that some of the famous canals pictured by the map actually exist. For example, the classical canal Agathodaemon, or Coprates (the only canal actually photographed from Earth), corresponds to the canyon named Valles Marineris. However, Valles Marineris is nearly as large as the United States, so it is more a grand-Grand Canyon than an aqueduct.

Perhaps the aspect of Edgar Rice Burroughs' map of Mars that interests Carl the most is not the map itself, but the way it was supposedly obtained. Edgar Rice Burroughs' hero, John Carter, traveled to Mars by simply lifting his arms. No long and lonely voyage or life support was needed. This idea sounds ridiculous, but truth being as strange as fiction, that is exactly what has been done over the past thirty years or so. An armada of robots has been sent by the nations of Earth to see, feel, and listen to the planets for us. During this spacecraft era we have peeked beneath the veil of mystery surrounding the planets. The planets are no longer mere figments of the imagination; they have become places.

A casual glance at a Viking lander image of Mars, like the one shown in Figure 5.1, suggests that Mars might not be such a bad place to visit. People who enjoy a desert lifestyle might even find these scenes to be reminiscent of home and think to themselves, "Here is a great place to take a stroll." However, such hasty judgments are dangerous. Before we step out on the red planet, we need to know what a visit to Mars actually would be like.

We may not be ready to send humans to Mars, but NASA's Ames Research Center and Arizona State University at Tempe operate a wind tunnel that simulates Martian conditions. Although it is large enough, you cannot stand inside the wind tunnel to see firsthand what Mars is like. However, you can peer through the window to see what the consequences of a walk on Mars might be. Some very interesting physical phenomena occur as the atmospheric pressure is reduced from terrestrial to Martian. A jar of water (imagine that this is the blood in your body) left standing inside the wind tunnel looks perfectly normal until pressures near those of Mars are reached. Then the water begins to boil furiously, but instead of hot steam it emits a frigid vapor that condenses to little pellets of ice. Within a few seconds the boiling water freezes solid into a jagged, porous lump of ice. The surface pressure on Mars is equivalent to that beyond an altitude of 100,000 feet above Earth's surface. If you ventured out to explore the Martian desert without a pressure suit, your body would literally explode as your blood boiled, killing you before you missed that first breath of oxygen that sustains us on Earth.

FIGURE 5.1
The surface of Mars as seen by the Viking-2 lander. The small furrows, next to the fallen part of the lander, were dug for experiments to determine the composition of the soil and to detect life.

I had another close encounter with Mars aboard NASA's DC-8 flying laboratory during the 1987 Airborne Antarctic Ozone expedition. Many planetary scientists go to the Antarctic coastline during summer to study terrestrial analogs to Mars. However, in order to learn the reasons for the sudden formation of the ozone hole we visited the atmosphere above the South Pole during midwinter – a time and place that are much more representative of Mars than is the summer-time Antarctic coastline. Surely this is the most isolated place on Earth. Seventeen men and one woman were living at the South Pole Station, which is located on a nearly featureless plateau and is cut off from the rest of the world through the long Antarctic winter.

As we flew past them on the spring equinox, I began to compare the environment around my aircraft to that of Mars. We were flying in the stratosphere through the ozone hole, so our atmosphere would do little to protect us from the ultraviolet light that effectively sterilizes the surface of Mars. Aurora overhead reminded me of the protection from high-energy cosmic rays that Earth's magnetic field and atmosphere offer. Outside air temperatures were so low, below 200 K, that the fuel had to be constantly circulated in the wings of the aircraft to keep it from freezing. Such temperatures, which are the norm on Mars, desiccate the atmosphere there. Likewise, Antarctica is a vast desert, whose reservoirs of ice have been supplied over hundreds of thousands of years of minimal precipitation. The atmosphere at the altitude of our aircraft contained just the same amount of water as does the atmosphere of Mars. There is enough water to fill the air with diamond-dust, small glittering ice crystals drifting downward everywhere through the frigid air, but hardly enough to maintain a decent cloud, let alone rivers and lakes.

The Antarctic environment is harsh, but it is much more conducive to life as we know it than is the Martian environment. Life has only a toehold at the South pole. Our lonely aircraft, containing fewer than 50 people, left behind a long contrail – such a definitive and unusual signal of life that when U.S. Air Force personnel observed it from space with a satellite, they sought the reason for it. We have seen no signal of life on Mars. There is no evidence for any large creatures, past or present. The Viking landers did not even observe any effects of Martian microbes on the chemistry of the soil or the atmosphere. The effects of such microbes are clearly evident on Earth. They maintain the oxygen in our atmosphere and pollute the air with other chemicals, such as methane, which are out of chemical equilibrium.

Ancient features on the surface of Mars (Figure 5.2) strongly resemble terrestrial river valleys, suggesting that Mars was once hotter and more humid than it is now. During the first 25% of Martian geologic history, an epoch of heavy bombardment of the Martian surface by asteroids and comets, water flowed across the Martian surface. The first

FIGURE 5.2
Nirgal Vallis:
the remnants of
an ancient,
800-kilometer-long
streambed in the
equatorial region
of Mars.

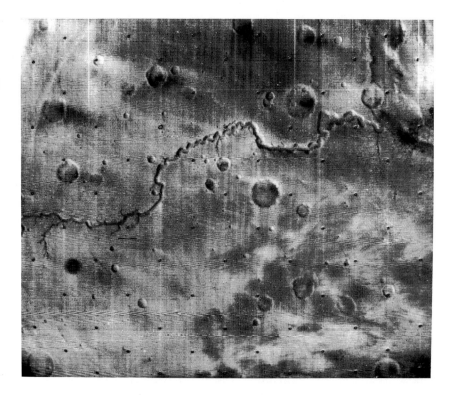

TABLE 5.1
TERRAFORMING MARS FOR PLANTS OR HUMANS

Parameter	Limits for Habitability	Mars Now
Global temperature	0–30°C	−60°C
Pressure		
Plants	>10 mbar	6 mbar
Humans	>500 mbar	
Nitrogen		
Plants	>1–10 mbar	0.2 mbar (atmosphere)
		2–300 mbar (regolith)
Carbon dioxide		
Plants	>0.15 mbar	10 mbar (atmosphere + poles)
Humans	<10 mbar	0.1–20 bar (regolith)
Oxygen		
Plants	>1 mbar	0.01 mbar (atmosphere)
Humans	>130 to <300 mbar	

fossils of simple organisms are found on Earth at the end of this time period. Perhaps one day we will find similar fossils on Mars. If the evolution of life on Mars followed the pace of terrestrial evolution, we might find fossil colonies of algae, but not the bones of dinosaurs.

Mars was once a better place for life. Perhaps it can be rehabilitated for humans or, failing that, at least for plants (Table 5.1). Even the fundamental requirements for life – liquid water, above freezing temperatures, and sufficient air pressure – are lacking on Mars. Water is almost certainly plentiful in the Martian soils, but the global temperature is too low (−60°C or about −75°F) for it to be in the liquid state. The atmospheric pressure is only slightly too low for plants, but much too low to support the water vapor pressures generated in warm-blooded humans.

There are also more subtle, but potentially more difficult, challenges to terraforming Mars. Nitrogen may be a life-limiting element on Mars. There is unlikely to be enough nitrogen, even in geologic reservoirs, to produce a terrestrial atmospheric pressure. Plants might find almost enough nitrogen in the current atmosphere for nitrogen fixation. However, if atmospheric nitrogen represents the complete supply of nitrogen, then only a very limited amount of biomass can be created. The Martian atmosphere contains enough nitrogen to supply the biomass density found on land and in the sea on Earth. Most of Earth's current biomass is present in the form of soil humus and decaying organic matter in the ocean; only about 6% is in living organisms. However, on Earth nearly as much nitrogen lies buried in sedimentary rocks – remnants of living organisms who passed on before us – as in our atmosphere. Therefore, the Martian atmospheric nitrogen cannot supply more than about 0.1% of the nitrogen contained in Earth's current and fossil biomass.

Carbon dioxide is often thought of as the key material for making Mars more habitable. Mars' atmospheric carbon dioxide supply is capable of maintaining a plant world. However, if we wished to reheat Mars using carbon dioxide as a greenhouse gas, we would have to access geologic reservoirs. Unfortunately, the resulting carbon dioxide atmosphere, which would have to have a total pressure exceeding Earth's to yield Martian temperatures near freezing, would be toxic to humans even if oxygen were present. In fact, current Mars' atmospheric carbon dioxide amounts are nearly toxic for humans. Greenhouse gases other than carbon dioxide will be needed if Mars is ever to be a home for unprotected humans.

Oxygen, a key requirement for some forms of terrestrial life, is present only in limited quantities in the current Martian atmosphere, but it is plentiful in the soils. The terrestrial oxygen atmosphere was created in the last 25% of Earth's history as a by-product of life. Oxygen availability does not limit our ability to terraform Mars.

Terraforming Mars would be a massive project. A synthetic atmosphere would have to be produced and constantly maintained. We can envision several ways to possibly achieve such an atmosphere, but lack of knowledge of the mineral reservoirs on Mars prevents us from determining if any of these schemes is practical. Instead of searching for gold, or black gold like terrestrial miners, astronaut prospectors may seek the currency of terraforming-accessible carbon and nitrogen reservoirs in the Martian soil.

Of course, planetary-scale terraforming is not the only way to make Mars habitable for humans. Astronauts may terraform small parts of Mars enclosed within protective shells, so that a human colony can be maintained, even if planetary-scale reengineering is just a distant dream.

Science fiction writers correctly captured the essence of the Martian environment in the first half of this century. They saw it as a vast desert, with a dwindling atmosphere. However, they envisioned our nearest neighbor, Venus, as a tropical paradise complete with humans and various prehistoric reptiles. But, Venus is more like Hades than the Garden of Eden.

If you were a visitor to Venus on an approaching spacecraft you might see the pale yellow clouds as a soft, welcoming haze with various interesting markings in the ultraviolet (Figure 5.3). However, a descent into the cloud tops, which extend nearly 70 kilometers above the surface, would reveal a mist of concentrated sulfuric acid that would badly burn your skin if you were exposed to it. These clouds are similar to stratospheric volcanic clouds on Earth. A visit to the cloud tops might remind you visually of a smoggy day in Los Angeles. As

FIGURE 5.3

To the eye, Venus is a nearly featureless, pale yellow disk. However, at ultraviolet wavelengths the clouds of Venus contain many interesting bright and dark features. The dark features are caused by the presence of a material that absorbs ultraviolet light; its composition remains unknown.

you descend closer to the surface you would enter a region of thicker, precipitating clouds. These clouds are similar to a dense fog in the lower atmosphere of Earth in which the sunlight becomes so diffuse that you can no longer locate the sun. Although these clouds resemble those on Earth, they drizzle nearly pure sulfuric acid, rather than a refreshing mist of water. Earth and Venus clouds differ markedly in composition. There is practically no water in the atmosphere of Venus from which to form water clouds. Earth, like Venus, has an active sulfur cycle, but sulfur emissions are quickly moved from the sky to the oceans of Earth, while they remain trapped in the atmosphere of Venus.

About 45 kilometers above the surface of Venus your spacecraft would break free from the clouds and enter clear sky with visibility similar to that of the cleaner regions of Earth. Now the pressure and

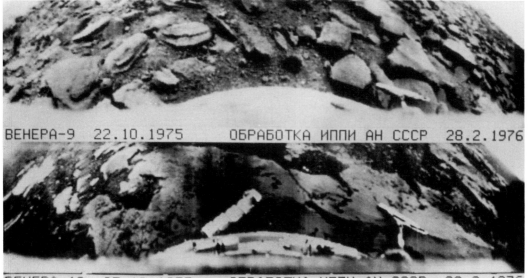

ВЕНЕРА-9 22.10.1975   ОБРАБОТКА ИППИ АН СССР 28.2.1976

ВЕНЕРА-10 25.10.1975   ОБРАБОТКА ИППИ АН СССР 28.2.1976

FIGURE 5.4
Views of the rocky
surface of Venus
near two Soviet
Venera landers.

temperature begin to build toward the surface, where you reach the
classic version of Hell (Figure 5.4). The surface pressure is equivalent
to being a thousand meters below the surface of a terrestrial ocean.
The temperature is that of an oven during its self-cleaning cycle. The
sunlit sky, thick with sulfurous gases, glows a rosy peach color, giving
the surface a strange warm tint, while the clouds block any view of the
sun or the heavens. Here is an environment so hostile that not even
our spacecraft can survive for more than a few hours.

Mars is too cold, dry, and airless to support life as we know it.
Venus is much too warm, and its atmosphere is too dense and cor-
rosive to be hospitable. Earth, to paraphrase Goldilocks, is just right.
Earth has remained just right for most of geologic history despite the
rising luminosity of our sun – the so-called faint young sun paradox
that Carl Sagan did much to call to our attention (Figure 5.5). The
dotted line in Figure 5.5 shows the rise in solar luminosity expected
for a solar-type star over geologic time. The solid line is the freezing
point of water. Without an atmosphere Earth should have temperatures
that follow the lower line labeled No Atmosphere; our current surface
temperature would be below freezing. However, the greenhouse effect
produced by our present atmosphere would cause Earth's temperature to have followed the curve labeled With Present Atmosphere.
Many questions arise from this figure. Following the With Present Atmosphere curve, why wasn't Earth a frozen block of ice in its early
history when the sun was dimmer? Because we know the Earth has
never frozen over, perhaps it followed some other temperature history
curve, but then why hasn't it superheated as the sun became brighter?

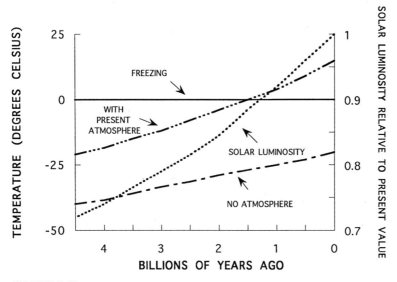

FIGURE 5.5

Description of the faint young sun paradox. As the solar luminosity rose through time, Earth's temperature should have increased. However, observations show that Earth has had roughly the same, above-freezing, temperature through geologic history. Without an atmosphere, the Earth would be much colder than it is with its present atmosphere. Could the atmospheric composition have been different in the first few billion years of geologic history than it is now, so that the surface temperature then exceeded the freezing point?

Because Earth is just right now, and has stayed that way over geologic time, does every planetary system in the vast universe have such a welcome abode for life? These questions remain unanswered, but several intriguing possible answers have been put forth.

The high surface temperature on Venus provides some clues toward answering these questions (Figure 5.6). At first one might imagine that the high temperature occurs because Venus is about 25% closer to the sun than is Earth. In fact, because the solar energy varies like distance from the sun squared, Venus receives nearly twice as much sunlight as Earth. However, the sulfuric acid clouds on Venus reflect much of the sunlight received by the planet back into space. If the positions of the planets are plotted considering the amounts of sunlight powering their climates, then, in fact, Venus is almost as far away from the sun as is frigid Mars. Therefore, Venus is not hot because of its proximity to the sun.

The Pioneer Venus mission measured the solar energy deposition profile and the major constituents of the atmosphere of Venus. Using these data we can calculate the vertical temperature profile of Venus and compare it with observations. These comparisons clearly show that the cause of the self-cleaning oven temperature found at the surface of Venus is the greenhouse effect that results from the immense amount of carbon dioxide in the atmosphere of Venus, which is about

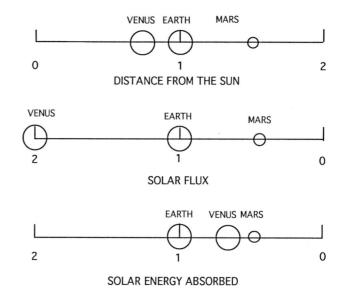

FIGURE 5.6
The relative positions of the planets are illustrated from three perspectives. At the top are the physical positions of the planets. In the middle are the relative amounts of solar energy available at the physical positions of the planets. At the bottom are the amounts of sunlight actually absorbed by the planets. At the bottom it can be seen that Venus, because of its highly reflective clouds, is effectively nearly as far from the sun as is Mars.

300,000 times the amount of carbon dioxide in the atmosphere of Earth.

Yet there is more to understanding the difference in surface temperatures between the Earth and Venus than knowing the current levels of atmospheric carbon dioxide. Earth has just as much carbon stored away in the form of limestones in crustal reservoirs as Venus has stored in the form of carbon dioxide in its atmosphere. Figure 5.7 shows that the storage of carbon in different places on Earth is related to liquid water. The arrows mark the path taken by terrestrial atmospheric carbon dioxide as it dissolves in fresh and salt water, along with calcium ions, which then are used by organisms to form limestone in the oceans. Without liquid water the carbon would build up in our atmosphere as carbon dioxide, and Earth would quickly become overheated like Venus.

Venus lost its water early in solar system history. Because of the high solar flux at the location of Venus, its original oceans of water vaporized into the atmosphere in a massive runaway greenhouse. At the top of the water-saturated atmosphere, water vapor dissociated into hydrogen and oxygen. The hydrogen escaped to space. So the position of Venus with respect to the sun is an important indirect factor in fixing its current temperature, because its position controlled the composition of the atmosphere.

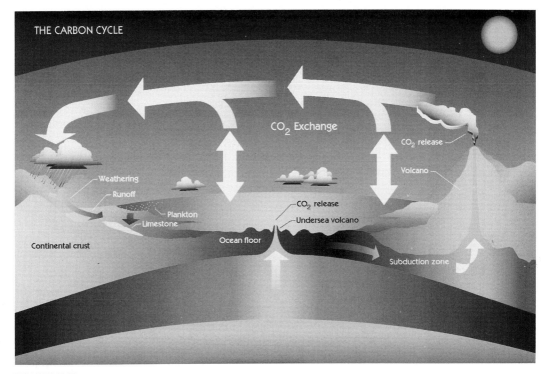

FIGURE 5.7

The geologic carbon cycle on Earth. Carbon dioxide from the atmosphere is slowly converted into carbonates on the sea floor due to the weathering of minerals from the continents and the construction of carbonate shells by microscopic organisms. Eventually the sea floor is carried beneath the continental crust by continental drift. There the sea floor minerals are heated until carbonates release carbon dioxide back to the atmosphere through volcanoes.

Mars provides further insight into the evolution of planetary atmospheres and climates. An atmospheric column on Mars contains more than 50 times the amount of carbon dioxide as an equivalent portion of Earth's atmosphere, yet it creates little greenhouse warming. Most of Earth's greenhouse warming is due to water vapor, and because Mars is so far from the sun it is too cold for much water vapor to be present in its atmosphere. For the surface temperature on Mars to increase to near the freezing point, so that water vapor could become abundant and produce a substantial additional greenhouse warming, the Martian atmosphere would need to contain about 300 times as much carbon dioxide as it does now.

Figure 5.7 also gives us clues to the fundamental reason that Mars is lacking the greenhouse gases that could keep it from being so cold. Early in its history, when liquid water was present on Mars, carbon dioxide would have reacted to form carbonates on Mars. Unlike Earth, however, Mars does not experience extensive volcanism. As can be seen in Figure 5.7, when sea floors are carried beneath the Earth's crust by continental drift, the limestone is heated to high temperatures and carbon dioxide is expelled through volcanoes. This volcanic source of carbon dioxide on Earth balances the loss due to carbonate formation in liquid water. On Mars the volcanic resupply of carbon dioxide does not occur, so if there were a denser, wetter atmosphere it would slowly vanish forever into the soils of Mars.

Why does continental drift and extensive volcanism not occur on Mars today? The answer may lie in the relatively small size of Mars – it is only twice as large as Earth's moon. Due to its small size, Mars soon lost its heat of formation and easily dissipates heat released from radioactive minerals, so the interior of Mars is now relatively cool. As Mars cooled, geological activity driven by thermal convection in its interior ceased, and with this loss of geologic activity the atmosphere of Mars was doomed.

The histories and fates of the three terrestrial planets suggest that a combination of distance from the sun, planetary size, as well as geologic and perhaps biologic evolution all contrive to control the habitability of the planets. These ideas can be extended to other solar systems. Zones of habitability have been drawn in Figure 5.8 for an Earth-sized planet in a stellar system with a solar-type star in an attempt to determine the likelihood of life elsewhere. Earth-like planets cannot remain habitable too much closer to a star than Earth's orbit, because the planets lose their water by photodissociation, and thick carbon dioxide atmospheres then build up. On the other hand, a planet might still be comfortable for life at the orbit of Mars, though it would

FIGURE 5.8
Calculated zones of habitability about a star like the sun. An Earth-like climate might be maintained over geologic time by an Earth-sized planet in the gray-shaded zone.

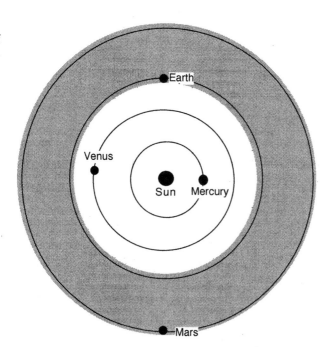

have to be large enough to have continental drift to maintain a relatively dense carbon dioxide atmosphere. Though such an atmosphere would be toxic to humans, its surface temperature could be moderate, liquid water could be present, and life forms adapted to this atmosphere might flourish. Surely the most exciting discovery of the next few decades will be the detection of planets within these habitable zones around other stars.

Venus, Mars, and Earth are the only planets on which spacecraft have landed so far, and of course we know the most about them because they are so close. Future explorations may reveal many other interesting habitats. What will it be like to visit Titan, which may have hydrocarbon seas? Do the organic compounds dripping from the sky there lead to any form of low-temperature life? Perhaps the Cassini mission with its Titan probe will tell us the answers. What would it be like to visit volcanically active Io, a sort of planetary Yellowstone? Do creatures live near the exploding sulfurous fountains, like they do in the sulfur pools and hydrothermal vents on Earth? Or do floating gasbags of life maneuver through the atmosphere of Jupiter? These questions may receive answers in the Galileo orbiter and probe missions. Cassini and Galileo will convert terra incognita to familiar ground.

During the spacecraft era the planets have become places. We may have no Michelin guide to the best hotels on Titan or to the restaurants on Mars. However, we have listed the major tourist spots. Like African explorers searching for the source of the Nile, we have inferred that glimmering hydrocarbon seas may exist on Titan. We have seen the Grand Canyon of Mars. In fact, in many hands-on museums children now navigate the canyons of Mars in computer-animated flyovers based on spacecraft images.

Fifty years ago armchair adventurers could explore the exotic domains of China or Earth's polar regions through the travelogs of human explorers, but they could only explore the planets with science fiction heroes like Edgar Rice Burroughs' John Carter. Today anyone can explore the planets using the world's planetary data and photographic archives or, better yet, by reading the many planetary travelogs written by Carl Sagan, who, surpassing John Carter, actually did voyage to the planets by lifting his arms.

6

The Origin of Life in a Cosmic Context

CHRISTOPHER F. CHYBA[*]

*White House Office of Science
and Technology Policy*[†]

The title of this essay is taken from a paper Carl Sagan published twenty years ago in the journal *Origins of Life* [Sagan 1974]. "The Origin of Life in a Cosmic Context" was one in a series of publications by Carl on the planetary context for life. Beginning in 1961 with "On the Origin and Planetary Distribution of Life" [Sagan 1961], he continued with the landmark book written with Iosef Shklovskii, *Intelligent Life in the Universe* [Shklovskii and Sagan 1966] and the article "Life" in the *Encyclopaedia Britannica* [Sagan 1970]. The time period of these articles spanned the entire U.S.–Soviet Moon race. But the 1974 paper was special in being the first in that series that could take into account the results of the lunar exploration by the Apollo astronauts.

The 1974 paper suggested three reasons why an extraterrestrial context was essential for understanding the origins of terrestrial life:

Distinguishing the Contingent from the Necessary. Only the study of extraterrestrial life will allow us to avoid the dangers inherent in trying to reach general conclusions about the nature of life based on the single example provided by terrestrial biology. "In our present profound ignorance of exobiology, life is a solipsism," Carl wrote. "There is no aspect of contemporary biology in which we can distinguish the evolutionary accident from the biological *sine qua non*. We cannot distinguish the contingent from the necessary." Does life really have to be based on proteins and DNA? Does it even have to be based on carbon? The danger of parochialism lurks in every conclusion we might reach.

[*]Current address: Department of Planetary Sciences, University of Arizona, Tucson, AZ 85721 USA.

[†]The ideas contained in this paper do not necessarily represent the views of the White House Office of Science and Technology Policy or the United States Government.

Determining the Timescale for the Origin of Life. How long did it take for life to evolve on Earth? Even if we can't yet answer this question, can we at least bound it? What is the longest that the origin of life could have taken – how long was the time period available for the origin of life on Earth?

Understanding the Origin of the Biogenic Elements. Elements such as carbon, nitrogen, and hydrogen make up the organic compounds from which terrestrial life is made. These biogenic elements had to be available on early Earth, and prebiotic organic compounds had to be synthesized in the absence of biology, prior to the origin of life.

Addressing each of these three issues illustrates why a cosmic context is valuable for making progress in understanding the origins of life on Earth. Indeed, one way of viewing progress in the field since 1974 – progress that has been especially substantial in the past decade (for a review, see [Chyba and McDonald 1995]) – is as progress in addressing these three issues.

The RNA and Other Possible Worlds

The most important recent advance in our understanding of the origins of life is the elucidation of a possible RNA world [Gilbert 1986, Joyce 1991]. The current world that we inhabit, of which the readers of this essay are exemplars, is a protein–DNA world. We metabolize using proteins, and we store our genetic information using DNA (deoxyribonucleic acid). DNA is needed to code for the structure of proteins, and proteins are in turn needed to produce DNA. As a result, there has been a long-standing chicken-or-egg paradox about the origins of life (see, e.g. [Dyson 1985]): if proteins are needed to make DNA and DNA is needed to make proteins, how could life have gotten started? Which came first, DNA or proteins?

However, beginning in the early 1980s, it was demonstrated that RNA (ribonucleic acid, a molecular cousin to DNA) was capable of significant enzymatic activity – that is, of performing the role of a protein. With the directed expansion of these abilities through test-tube evolution, it appears that RNA is capable of enough enzymatic power to have provided the basis of a primitive metabolism – while simultaneously serving as that system's genetic material (e.g. [Cech 1993]).

The elaboration of the RNA world concept suggests that we have, in fact, made progress in distinguishing the evolutionarily contingent from the biologically necessary – in the absence of an exobiological system for comparison. It appears we've learned that there is at least one other kind of possible life, RNA life. This conclusion has been made plausible both by unearthing largely hidden capabilities of contemporary RNA molecules and by expanding the capabilities of those molecules through directed molecular evolution in the laboratory. Protein–DNA life is apparently not the only possible form of life.

It is the assumption that extraterrestrial biology would be based on chemistry that allows some progress to be made even in the absence of full-blown biological examples. Consider the old exobiological speculation of life based on silicon, rather than carbon. In his 1961 paper, Carl argued "silicates lack the information-carrying properties of variable side chains which characterize such carbon compounds as polynucleotides and polypeptides. Therefore it is doubtful that silicates could be a fundamental constituent of extraterrestrial organisms."

So it seems that we can, and have, made some progress toward understanding alternate possibilities for life, even in the absence of examples of extraterrestrial life. The RNA world shares many characteristics of our DNA–protein world, so it does not test the limits of the possible as far as might extraterrestrial life. Nevertheless, Carl may have been too pessimistic in his view of what we could hope to accomplish without an extraterrestrial example to guide and test us.

Timescales for the Origin of Life

Our understanding of the length of time available for the origin of life on Earth has been changed through the exploration of the Moon by U.S. and Soviet spacecraft and the return of samples to Earth. This exploration made it clear that to understand the origins of life on Earth, we have to go to the Moon and the planets. This isn't just a metaphor, but rather is literally true.

Microscopic fossils, and fossils of microbial mats, or stromatolites, have been found in terrestrial sedimentary rocks that are 3.5 billion years (Gyr) old. The Earth is almost 4.6 Gyr old. The ocean floors are, by contrast, very young, typically less than 200 million years (Myr) old. Most continental crust postdates the Archean eon, which ended 2.5 Gyr ago. And most of the Archean crust is younger than 3.5 Gyr. In fact, only in western Australia, southern Africa, and southern Greenland are there sedimentary rocks as old as 3.5 Gyr. The western Australian and southern African terrains date from about 3.5 Gyr ago. The Isua formation in Greenland consists of 3.8-Gyr-old metasediments. The Earth is so geologically active that, as far as we know, there aren't any older sedimentary or metasedimentary rocks left anywhere else on the planet [Veizer 1983]. This is of special interest because only rocks originating as sediments can contain fossils.

The fascinating thing about each of these most ancient terrains is that all three preserve signs of ancient life. There are unambiguous 3.5-Gyr-old fossils in western Australia and likely candidates in southern Africa [Schopf and Walter 1983, Schopf 1993]. The Isua metasediments are sediments that have subsequently experienced high temperatures and pressures during their long history, so they appear incapable of having retained fossils. Nevertheless, the carbon isotope ratio ($^{13}C/^{12}C$) in Isua organic carbon may be indicative of isotopic fractionation by photosynthesizing life [Schidlowski 1988] – though this conclusion remains controversial.*

So the oldest geological evidence for conditions on the early Earth shows signs of life – and perhaps sophisticated, photosynthesizing life. This makes it very difficult to understand the nature of the terrestrial environment at the time of the origins of life by looking at geological evidence. By the time of the oldest evidence we have, it's already too late – life already exists. We cannot learn about conditions on prebiotic Earth by looking at the Earth, because the slate has been wiped clean. But we can learn some important things about the prebiotic Earth by looking at the Moon.

Results of the lunar landings showed that most of the surface of the Moon is very ancient; much of it is older than the oldest sedimentary rocks on Earth. Thus, the Moon provides a window to the first billion years of Solar System history. The Moon has been geologically dead for three billion years so, unlike the Earth, it has retained much of the record of this early time.

Radioactive dating of returned lunar samples allows the reconstruction of a history of lunar cratering. Dating rocks returned from different lunar terrains, combined with crater counts for those terrains [Basaltic Volcanism Study Project 1981], yields a plot of crater counts per unit lunar area as a function of time (e.g. [Chyba, Owen, and Ip 1995]). This plot suggests that, although lunar cratering over the past 3.5 Gyr has been at a relatively low level, comparable to that of today, cratering for the first billion years of lunar history was exponentially higher. Knowing how much greater Earth's gravity is than that of the Moon, as well as typical velocities for impacting projectiles, it is straightforward to extrapolate the lunar cratering history to the Earth. We can therefore make statistically reasonable estimates of Earth's cratering history for its first billion years, even though little terrestrial record of that time remains. The Moon is critical to an accurate general picture of early Earth's impact environment.

Why is this important? Everyone is now familiar with the idea that a comet or asteroid collided with the Earth 65 Myr ago, at the end of the Cretaceous period [Alvarez et al. 1980]. (The geological period immediately following the Cretaceous was the Tertiary; hence, the impactor is called the Cretaceous–Tertiary, or K/T, impactor.) That collision,

---

*Note added in proof: Since this chapter was written, the isotopic evidence for life 3.85 Gyr ago (from rocks from Akilia island, Greenland) has become much stronger. See Mojzsis, S. J., Arrhenius, G., McKeegan, K. D., Harrison, T. M., Nutman, A. P., Friend, C. R. L. 1996. Evidence for life on Earth before 3,800 million years ago. *Nature* 384:55–59.

which excavated the 200-kilometer diameter Chicxulub crater [Sharpton et al. 1992], was the impact of an object about the size of Halley's comet – ten kilometers or so in diameter. It now seems quite likely that the K/T impact had a profound influence on the history of life on Earth. The fact that we are here – rather than, say, six-foot green reptiles – may be the result of this impact. By extrapolating the lunar cratering record to the Earth, it is clear that objects the size of the K/T impactor collided with Earth some 10,000 times during its first billion years. At the time of the origins of life, say about 4 Gyr ago, Earth was sustaining such an impact every few hundred thousand years.

The lunar cratering record also reveals that the flux of impacting bodies increased in number as something like the 1.6 power of decreasing impactor diameter. Therefore, for example, Earth 4 Gyr ago would also have been experiencing the impact of a 1-kilometer diameter body every 5,000 years or so.

Clearly, this was an extremely hostile environment. Darwin, in his famous 1871 letter to Hooker (reproduced, e.g., in [Hartman, Lawless, and Morrison 1985]), imagined that life had originated in some warm little pond. This is a peaceful image, leading one to envision quiet, serene conditions. But the Moon teaches us that this is not the right picture. The right picture is that life on Earth must have originated not in quiet, peaceful circumstances, but rather in extremely violent, impact-ridden ones.

The largest craters on the Moon – big, multiringed structures greater than 300 kilometers across – are called basins. We now know of about fifty such lunar basins [Wilhelms 1987], ranging in size up to the South-Pole Aitken basin, about 2,200 kilometers in diameter [Belton et al. 1992]. The largest basin that one can see on the lunar near side is Imbrium, about 1,160 kilometers across. Objects with radii in the range 50–150 kilometers are required to excavate basins of this size. With a terrestrial gravitational cross section nearly 25 times bigger than that of the Moon, and about 50 known lunar basins, it's clear Earth sustained a large number of giant impacts.

The most gigantic of these giant impacts would have had extremely grave consequences for life on Earth. Work by Maher and Stevenson [1988] and Sleep et al. [1989] suggests that the biggest might well have sterilized the surface of the Earth. In this case, the window available for the origin of life was only as wide as the time between the oldest geological evidence for life (either 3.5 or 3.8 Gyr ago) and the final sterilizing impact. With the heavy bombardment of the Moon not tailing off until around 3.8 Gyr ago (the Imbrium impact, incidentally, seems to have occurred 3.85 Gyr ago), the timescale for the origin of life on Earth shrinks from nearly a billion years down to around 100 Myr, or even less. The suggestion that the lunar bombardment evident in the Apollo results implied a narrower window for the origin of life than we had previously supposed, probably only around 100 Myr long, was to my knowledge first made by Carl in his 1974 paper.

What happens when the Earth gets struck by one of the biggest of the big impactors? Sleep et al. [1989] examined this question in detail, modeling the aftermath of the collision with Earth of an object about 400 kilometers in diameter – roughly the size of the large asteroids Vesta, Pallas, or Hygiea. Early Earth probably sustained a handful of such collisions. In impacts of this magnitude, the collision creates a rock-vapor atmosphere that encircles the globe and persists for several months. So for a period of several months, if you were an organism at the terrestrial surface, you would be breathing rock vapor. This is bad – bad not only for full-blown organisms but also for prebiotic chemistry as well. And it gets worse. This rock vapor atmosphere ablates the ocean, evaporating it entirely. For a period of a few thousand years, Earth would as a result be enveloped in a dense steam atmosphere that bakes the terrestrial surface to the depth of a few hundred meters to about a thousand degrees kelvin. By the end of this time, the ocean would have rained back down to the surface.

Arguably this is a planet-sterilizing event. It seems plausible that every time a collision of this magnitude occurs, whatever promising experiments in the origin of life may be underway are annihilated, and prebiotic chemistry must begin again from scratch. The clock is reset. From this point of view, the fact that life on Earth today is based on our particular biochemistry may be a kind of stochastic result. Had the last Earth-sterilizing impact missed us, or had there been yet one more after our ancestors finally took hold 3.8 to 3.5 Gyr ago, contemporary life on Earth might itself be an example of one of the radically different biologies to which Carl was looking in 1974 for greater perspective.

Nevertheless, it is not certain that one of these giant impacts would necessarily have sterilized the planet. Heating the outer hundreds of meters of the Earth to a thousand degrees would not be equivalent to sterilizing the Earth if life had originated and had time to migrate into protected niches at depths sufficient to remain insensitive to the surface heating. It is of great interest in this regard that there are con-temporary terrestrial bacteria known to flourish at depths of several kilometers [Boston, Ivanov, and McKay 1992]. In the event that Earth were home to a deep, hot biosphere in the terrestrial crust [Gold 1992], especially if one envisions life as having originated at such depths, even Vesta-scale impacts might be of little bother. A much greater un-derstanding of subsurface terrestrial life is needed before we know how to assess this aspect of the problem.

In addition to the most giant impacts, there must have been many more that would have evaporated only some fraction of the oceans. An example would be those impacts that evaporated the upper 200 meters of the ocean, the photic zone, where photosynthesizing life is possible.

Statistically, it is likely that one or more photic-zone-destroying impacts occurred subsequent to the final ocean-evaporating impact, so the ancestors of contemporary terrestrial life must have surmounted such global catastrophes. It may have been that the only life that survived was life that had remained down in, or evolved down to, the depths.

It has been suggested that the early ancestors of extant microorganisms were thermophiles growing near the temperature of boiling water (e.g. [Woese 1987]). This could mean either that the first organisms on Earth evolved at depth (and at the corresponding high temperatures) or, instead, that they evolved at the surface, but only those that had spread quickly into deep protected niches made it through the impact bottleneck.

Impact Delivery of Biogenic Elements and Prebiotic Organics

So far I've been discussing the potential devastation wrought by cosmic collisions with the Earth. From this point of view, impacts would have shortened the time available for the origin of life on Earth. Yet, there is a sense in which impacts could also have helped in the origin of life.

Comets and many asteroids are extremely rich in the biogenic elements, those volatile elements like carbon and nitrogen that are essential to organic life. Halley's Comet, for example, is about 25% organic by mass. That might seem incredible, but this is what is implied by comets' roughly cosmic abundances of these elements, and it is borne out by spacecraft measurements (e.g., [Chyba et al. 1990, Delsemme 1991]).

Earth, on the other hand, is extremely deficient in such elements. Many terrestrial formation models suggest Earth would have formed extremely poor in volatiles (e.g. [Delsemme 1992]). It appears possible, if not likely, that whatever biogenic elements Earth does have were brought to it subsequent to its bulk formation as a kind of impact veneer from collisions of comets and carbonaceous asteroids. In this picture, comets and some asteroids would have served as a kind of conveyor of volatile elements (including biogenic elements) from the outer Solar System, where those elements are abundant, to the inner Solar System, where temperatures were high enough for liquid water – the sine qua non of life as we know it – but where biogenic elements were extremely rare.

Quantitatively, using the lunar cratering record to estimate the total mass accreted by Earth subsequent to 4.4 Gyr and prior to 3.5 Gyr ago, it turns out that if some tens of percent by mass of the objects that excavated the lunar basins and craters were cometary, Earth would have accreted about an ocean's worth of water. It now seems plausible that Earth collected virtually its entire complement of water, carbon, nitrogen, and other biologically critical compounds and elements in this way [Chyba et al. 1995].

Finally, one can ask whether Earth's early primordial soup might have been stocked with prebiotic organics in this same manner – via impact delivery during the heavy bombardment. This is difficult to do. If an object the size of Halley's Comet, for example, were to hit the Earth, the resulting explosion would have an energy of about a hundred million megatons. It seems prima facie unlikely that organic molecules could survive such conditions. Indeed, a few years ago, working with our colleagues Paul Thomas and Leigh Brookshaw, Carl Sagan and I modeled the survival of organics in cometary and asteroidal impacts with Earth, and found that exceptional circumstances were necessary to have organics survive [Chyba et al. 1990, Thomas and Brookshaw 1997].

However, there is a startling datum from the K/T boundary. The clay layer at the K/T boundary is of course rich in iridium; it was this iridium anomaly that first led Alvarez et al. [1980] to posit a giant impact as the extinction trigger. Remarkably, it appears that the boundary is also rich in the amino acids α-amino isobutyric acid and racemic isovaline [Zhao and Bada 1989]. These amino acids are extremely rare in the biosphere, but are among the most common in meteorites and in laboratory prebiotic synthesis experiments. Their discovery at the boundary suggests that there is a mechanism, perhaps quench synthesis in an expanding fireball, to produce large quantities of interesting prebiotic organics in big impacts. Indeed, in some likely early terrestrial atmospheres, exogenous sources of organics may have exceeded endogenous atmospheric production from ultraviolet light or electrical discharges [Chyba and Sagan 1992, 1997].

Panspermia and Mars

To conclude, I'd like to mention one peril of hoping to gain a cosmic context for terrestrial life by searching for extinct (presumably microscopic) life on Mars. Although the difficulties faced by traditional schemes for panspermia across interstellar distances are well known [Davies 1988], it is possible that microorganisms could successfully migrate between Earth and Mars in the debris ejected from the vicinity of giant impacts [Melosh 1988]. Obviously this kind of exchange would have been most likely during the time of the heavy bombardment. We know that we have pieces of Mars in our museums, in the form of the Shergottite-Nakhlite-Chassignite (SNC) meteorites, and that some of these show evidence for very little shock heating. The physics of planetary ejection in the absence of high shock pressures and temperatures is now well understood [Melosh 1989].

The SNC meteorites have a great deal to tell us about the possibility of an ancient Martian hydrosphere [Karlsson et al. 1992] and Martian organics [McDonald and Bada 1995]. Bigger pieces of Mars, ten meters or more in size, could shield microorganisms from cosmic rays during interplanetary passage, so that it's possible that microorganisms may have successfully migrated between Mars and Earth early in these planets' histories.

In the future, as we explore Mars, and as we look for signs of an ancient Martian biosphere, it will be precisely those fossils that are easiest to recognize as fossils – because of their similarities to terrestrial microfossils – for which it will be the hardest to tell whether they represent a separate Martian origin of life or merely resulted from a kind of inoculation of Mars by the Earth. (For that matter, they could even represent the now-extinct ancestors of terrestrial life, whose only living descendants are the progeny of those that survived the perilous journey to Earth.) Would the discovery of microscopic fossils that resembled terrestrial microorganisms tell us that life on other worlds must find the same solutions to the problems of early evolution? Or would it merely mean that the two worlds had exchanged organisms early in their history? It is not clear that fossil discoveries on Mars will necessarily help in providing the kind of perspective that Carl has asked for. That kind of perspective may be even more difficult to attain than we had hoped.*

BIBLIOGRAPHY

Alvarez, L. W., Alvarez, W. A., Asaro, F., Michel, H. V. 1980. Extraterrestrial cause for the Cretaceous–Tertiary extinction. *Science* 208:1095–1108.
Basaltic Volcanism Study Project 1981. 1981. *Basaltic Volcanism on the Terrestrial Planets.* New York: Pergamon Press.
Belton, M. J. S., Head, J. W., Pieters, C. M., Greeley, R., McEwen, A. S., Neukum, G., Klaasen, K. P., Anger, C. D., Carr, M. H., Chapman, C. R., Davies, N. E., Fanale, F. P., Gierasch, P. J., Greenberg, R., Ingersoll, A. P., Johnson, T., Paczkowski, B., Pilcher, C. B., Veverka, J. 1992. Lunar impact basins and crustal heterogeneity: New western limb and far side data from Galileo. *Science* 255:570–576.
Boston, P. J., Ivanov, M. V., McKay, C. P. 1992. On the possibility of chemosynthetic ecosystems in subsurface habitats on Mars. *Icarus* 95:300–308.
Cech, T. R. 1993. The efficiency and versatility of catalytic RNA: Implications for an RNA world. *Gene* 135:33–36.
Chyba, C. F., McDonald, G. D. 1995. The origin of life in the Solar System: Current issues. *Annu. Rev. Earth Planet Sci.* 24:215–249.

*Note added in proof: Since this chapter was written, possible microfossils and other possible signs of life have been reported in the Martian meteorite ALH84001. These claims remain controversial. See McKay, D. S., Gibson, E. K., Thomas-Keprta, K. L., Vali, H., Romanek, C. S., Clemett, S. J., Chillier, X. D. F., Maechling, C. R., Zare, R. N. 1996. Search for past life on Mars: Possible relic biogenic activity in Martian meteorite ALH84001. *Science* 273:924–930.

Chyba, C., Sagan, C. 1992. Endogenous production, exogenous delivery and impact-shock synthesis of organic molecules: An inventory for the origins of life. *Nature* 355:125–132.

Chyba, C. F., Sagan, C. 1997. Comets as a source of prebiotic organic molecules for the early Earth. *Comets and the Origin and Evolution of Life.* In P. J. Thomas, C. F. Chyba, C. P. McKay, eds., New York: Springer-Verlag, 147–173.

Chyba, C. F., Thomas, P. J., Brookshaw, L., Sagan, C. 1990. Cometary delivery of organic molecules to the early Earth. *Science* 249:366–373.

Chyba, C. F., Owen, T. C., Ip, W.-H. 1995. Impact delivery of volatiles and organic molecules to Earth. In T. Gehrels, ed., *Hazards Due to Comets and Asteroids.* Tucson: University of Arizona Press, 9–58.

Davies, R. E. 1988. Panspermia: Unlikely, unsupported, but just possible. *Acta Astron.* 17:129–135.

Delsemme, A. H. 1991. Nature and history of the organic compounds in comets: An astrophysical view. In R. I. Newburn, M. Neugebauer, J. Rahe, eds., *Comets in the Post-Halley Era.* Dordrecht: Kluwer, 377–428.

Delsemme, A. H. 1992. Cometary origin of carbon, nitrogen and water on the Earth. *Orig. Life Evol. Biosph.* 21:279–298.

Dyson, F. 1985. *Origins of Life.* Cambridge: Cambridge University Press.

Gilbert, W. 1986. The RNA world. *Nature* 319:618.

Gold, T. 1992. The deep, hot biosphere. *Proc. Natl. Acad. Sci. USA* 89:6045–6049.

Hartman H., Lawless, J. G., Morrison, P., eds. 1985. *Search for the Universal Ancestors.* Washington: NASA SP-477.

Joyce, G. F. 1991. The rise and fall of the RNA world. *New Biol.* 3:399–407.

Karlsson, H. R., Clayton, R. N., Gibson, E. K., Mayeda, T. K. 1992. Water in SNC meteorites: Evidence for a martian hydrosphere. *Science* 255:1409–1411.

Maher, K. A., Stevenson, D. J. 1988. Impact frustration and the origin of life. *Nature* 331:612–614.

McDonald, G. D., Bada, J. L. 1995. A search for endogenous amino acids in the martian meteorite EETA 79001. *Geochim. Cosmochim. Acta* 59:1179–1184.

Melosh, H. J. 1988. The rocky road to panspermia. *Nature* 363:498–499.

Melosh, H. J. 1989. *Impact Cratering: A Geologic Process.* New York: Oxford University Press.

Sagan, C. 1961. On the origin and planetary distribution of life. *Radiat. Res.* 15:174–192.

Sagan, C. 1970. Life. In *Encyclopaedia Britannica.*

Sagan, C. 1974. The origin of life in a cosmic context. *Orig. Life Evol. Biosph.* 5:497–505.

Schidlowski, M. A. 1988. 3,800-million-year isotope record of life from carbon in sedimentary rocks. *Nature* 333:313–318.

Schopf, J. W. 1993. Microfossils of the early Archean apex chert: New evidence for the antiquity of life. *Science* 260:640–646.

Schopf, J. W., Walter, M. R. 1983. Archean microfossils: New evidence of ancient microbes. In J. W. Schopf, ed., *Earth's Earliest Biosphere: Its Origin and Evolution.* Princeton: Princeton University Press, 214–239.

Sharpton, V. L., Dalrymple, G. B., Marin, L. E., Ryder, G., Schuraytz, B. C., Urrutia-Fucugauchi, J. 1992. New links between the Chicxulub impact structure and the Cretaceous/Tertiary boundary. *Nature* 359:819–821.

Shklovskii, I. S., Sagan, C. 1966. *Intelligent Life in the Universe.* New York: Dell.

Sleep, N. H., Zahnle, K. J., Kasting, J. F., Morowitz, H. J. 1989. Annihilation of ecosystems by large asteroid impacts on the early Earth. *Nature* 342:139–142.

Thomas, P. J., Brookshaw, L. 1997. Numerical models of comet and asteroid impacts. In P. J. Thomas, C. F. Chyba, eds., McKay, C. P., *Comets and the Origin and Evolution of Life.* New York: Springer-Verlag, 131–145.

Veizer, J. 1983. Geologic evolution of the archean-early proterozoic Earth. In J. W. Schopf, ed., *Earth's Earliest Biosphere.* Princeton: Princeton University Press, 240–259.

Wilhelms, D. E. 1987. *The Geologic History of the Moon.* Washington: U.S. Geological Survey SP 1348.

Woese, D. R. 1987. Bacterial evolution. *Microbio. Rev.* 51:221–271.

Zhao, M., Bada, J. L. 1989. Extraterrestrial amino acids in Cretaceous/Tertiary boundary sediments at Stevns Klint, Denmark. *Nature* 339:463–465.

7

Impacts and Life: Living in a Risky Planetary System

DAVID MORRISON

NASA Ames Research Center,
Space Science Division

In July 1994 the inhabitants of planet Earth witnessed an interplanetary collision. Ever since the time of Newton most scientists had thought of the solar system as a deterministic machine, obeying the laws of nature with clockwork precision. There was little room in our cosmology for violent collisions. To be sure, the cratered surfaces of the Moon and planets testify to collisions early in the history of the Solar System, but most people considered that such events were either restricted to ancient times when the planets were young or happened so infrequently that they were of no contemporary concern. Yet, we found ourselves in the summer of 1994 watching just such an event as the fragments of Comet Shoemaker–Levy 9 (S–L 9) plunged one after another into the atmosphere of Jupiter. This unprecedented series of impacts erased forever the illusion that the solar system is free from catastrophic collisions.

The S–L 9 impacts with Jupiter came at a time when fundamental changes were taking place in our conception of the role of cosmic collisions. Over the previous 30 years our understanding of the role of collisions had gradually emerged as spacecraft explored planetary surfaces throughout the solar system, demonstrating that crater-forming impacts were a dominant force in planetary history. The same three decades revealed the presence of more than 150 impact craters on Earth, scars from the continuing bombardment of our planet. However, the most important contribution to this revolution occurred in 1981, when Luis and Walter Alvarez and their colleagues (University of California at Berkeley) discovered that an extraterrestrial impact had been responsible for the mass extinction that took place 65 million years ago, at the end of the Cretaceous period of geological history. Until then, impacts and the craters they produce were of interest only to geologists. What the Alvarez team discovered was that even relatively modest impacts can modify the environment so severely that the course of biological evolution is profoundly altered. The history of life on this planet, and by implication the origin of humans, is thus closely coupled to the impact history of the Solar System.

This paper deals with a new paradigm: that life evolved on Earth in an environment punctuated by impact catastrophes. We have learned that our planet lives in a "bad neighborhood," with occasional outbursts of violence that have dramatically influenced our history. We are particularly interested in the contemporary hazard imposed by impacts and with proposals for ways to deal with these risks.

Carl Sagan has played an important role in the revolution we are discussing. As the world's foremost exponent of planetary exploration, he has done much to stimulate missions to the planets, to plan the critical science that was done on those missions, and to interpret the results for the broadest possible audience. Carl has also played a leading role in recognizing the importance of impacts and the existence of a contemporary threat. And he has been especially sensitive to the geopolitical implications of the impact hazard and the impetus this threat has given to possible defense schemes, many of which are potentially more dangerous than the natural hazard they are designed to mitigate. His unique combination of technical expertise and humane wisdom will be important for the public policy debates that will grow around the impact hazard and potential defensive systems.

Lessons from the Comet Crash

When Comet S–L 9 was discovered, some months after it had broken apart during a very close encounter with Jupiter, it already consisted of more than 20 fragments, each in orbit about Jupiter (Plate VIII). As a result of perturbations introduced by solar gravity, these orbits were distorted to yield a planetary collision the next time around. Over a period of about a week in July 1994, each of the fragments was doomed to impact the planet at 44° south latitude, just over the horizon as seen from the Earth. Although terrestrial telescopes could not image the actual entry and explosion from each impact, they would be well placed to see any large plume of material ejected above the impact site. And within a few minutes of each event, the impact site would be carried into direct view by the rapid rotation of the planet.

In the months preceding the impacts, a number of models were developed to predict the consequences as material smashed into the planet at a speed of 70 kilometers/second, more than three times the reentry speed of the Apollo lunar capsules. The stress of deceleration in the Jovian atmosphere would lead to the rapid breakup of each impactor. Unfortunately, however, the sizes of the fragments could not be measured, leading to widely divergent predictions for the observable phenomena. Models of the tidal breakup of the comet during its 1992 flyby of Jupiter suggested that the largest fragments were no more than 1 kilometer in diameter. On the eve of the impacts, many (perhaps most) observers feared that the comets would be swallowed by Jupiter without a visible trace.

These fears were not realized, and even the smaller of the S–L 9 impacts produced striking consequences, with direct implications for the terrestrial impact hazard. Each of the larger fragments apparently penetrated to a depth that did not quite reach the Jovian water clouds, where it disintegrated releasing a total energy of approximately 100,000 megatons – considerably larger than the Earth's entire inventory of nuclear weapons. Much of this energy was directed upward along the path of the incoming meteor, to produce a plume of hot gas that rose to an altitude of nearly 4,000 kilometers above the upper cloud level.

Since the impact plumes were moving too slowly to escape from Jupiter, they soon collapsed. Any dust or other solid material in the plumes reentered the atmosphere at speeds of the order of 10 kilometers/second and flashed into incandescence. The resulting intense meteor shower heated the gas and dust and lit the skies of Jupiter with infrared radiation. Since by this time the impact site had rotated into view, the infrared pulse was widely observed by terrestrial astronomers. An analogous meteor storm would occur on Earth following a large impact, but in our case the heat pulse would be distributed over most of the planet, igniting forests and grassland on a global scale. It is just such a conflagration that probably accounted for the extinction of the dinosaurs and many other species of terrestrial flora and fauna 65 million years ago.

The plume material that cascaded back into the Jovian atmosphere remained suspended in the stratosphere to produce extensive dark clouds that looked like smudges or bruises on the face of the planet (Plate IX). These dark clouds were easily seen in even small amateur telescopes, and they became the most prominent features on the disk. Typically, these dark clouds were about 15,000 kilometers across, larger than the planet Earth. In a similar way, terrestrial impacts would produce dark stratospheric clouds. However, in the case of an impact on our planet, a much larger fraction of the energy is channeled into dust production as the impactor smashes into solid ground. It is estimated that the dust cloud following a major impact would plunge the entire surface into profound darkness, bringing freezing temperatures on land and a breakdown in the oceanic food chain. Presumably it was the dust cloud that led to the marine deaths that define the mass extinction that terminated the Cretaceous period. On Jupiter, the dark clouds persisted for months, and the same would be expected on the Earth.

A third observation of Jupiter bears on the terrestrial effects of impacts. Surrounding each of the larger impacts, astronomers measured outward spreading atmospheric waves. Analogous but much more destructive waves occur on Earth if the impact takes place in the ocean. If an object with the energy of one of the S–L 9 fragments hit in the North Atlantic, tsunami more than 100 meters high would strike both North America and Europe, devastating the coasts and obliterating many of the world's great cities.

The observations of S–L 9 thus exemplify a number of the catastrophic phenomena that accompany impacts. These include a blast plume, a heat pulse generated by the plume debris as it reenters the atmosphere about 20 minutes after the impact, a long-lived stratospheric dust cloud, and waves (tsunami) that can travel for many thousands of kilometers from the impact site. None of these impacts did any damage to Jupiter as a whole; its orbit and rotation were unaffected. Yet, important changes took place in its atmosphere, producing new features larger than the planet Earth. In a similar way, impacts of kilometer-scale objects on the Earth are of no consequence for the orbit or the solid surface of the planet, but they can produce transient environmental changes that are catastrophic for life. It is the sensitivity of the atmosphere to impacts, and the sensitivity of life to changes in the atmosphere, that shift the impact phenomenon from the purely academic to the public policy domains.

Terrestrial Impacts

Both comets and asteroids are capable of striking the Earth. The first step in analyzing cosmic impacts is to determine numbers of such impacts. The average total impact flux over the past 3 billion years can be found from the crater density on the lunar maria, which were formed by widespread volcanism that erased previous features and created a clean slate for recording impacts. The Earth is subject to the same impact flux as the Moon, although our atmosphere filters out the smaller projectiles. We can also estimate the contemporary impact rate on the Earth from a census of existing Earth-crossing asteroids and comets, yielding a result that is consistent with the average rate recorded on the Moon. For most purposes it does not matter whether the impact is from a comet or asteroid; what counts is the power of the blow, not the composition of the hammer. Figure 7.1 summarizes the average terrestrial impact flux as a function of projectile kinetic energy, measured in megatons (1 megaton = energy of one million tons of TNT [trinitrotoluene] = 4.2×10^{15} joules).

On the basis of the average flux of comets and asteroids striking the Earth, we can evaluate the danger posed by impacts of different magnitudes. Of particular interest are two threshold sizes: the threshold for penetration though the atmosphere, and the threshold at which impacts have major global as well as local effects.

The atmosphere protects us from smaller projectiles. Figure 7.1 indicates that an impact event with the energy of the Hiroshima nuclear bomb occurs roughly annually, whereas a megaton event is expected

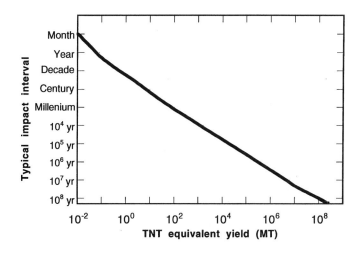

FIGURE 7.1

Plot of the long-term average frequency of impacts of different energy on the Earth as derived from studies of the lunar cratering record and observations of contemporary near-Earth asteroids and comets. The line shows the average interval between impacts of a given size (measured in megatons of energy) or larger for the entire area of the planet.

at least once per century. Obviously, however, such relatively common meteoric explosions are not destroying cities or killing people. Even at megaton energies, most projectiles break up and are consumed before they reach the lower atmosphere. For most impacting bodies, the threshold diameter to reach the surface is about 100 meters (the size of a football field), and the threshold for penetration to the lower atmosphere is about 50 meters. Objects smaller than this pose no danger, although their high-altitude explosions are routinely detected by military surveillance satellites; the largest reported bolide exploded 21 kilometers above the Western Pacific Ocean on February 1, 1994, with an estimated energy equal to several Hiroshima-scale bombs.

If the projectile is large enough and strong enough to penetrate below about 15 kilometers altitude before it explodes, the resulting airburst can be highly destructive. A historical example is provided by the Tunguska event of 1908, in which a stony asteroid about 60 meters in diameter penetrated to within 8 kilometers of the surface before exploding. The yield of the Tunguska blast has been estimated at 15 megatons, and it destroyed an area of 2,000 square kilometers. If such an impact took place over a heavily populated area, the results would be catastrophic. However, a Tunguska-yield impact takes place over the land area of the Earth only about once per millennium, and there is no historical example of the destruction of a city by such an impact. Obviously, the risk from such events is much lower than that from better-known natural disasters such as earthquakes and severe storms, each of which destroys or badly damages several cities somewhere on Earth within the span of a human lifetime.

At sufficiently great energies, an impact has global consequences. An obvious, if extreme, example is the Cretaceous impact of 65 million years ago, which disrupted the global ecosystem and led to a major mass extinction. This impact of a 15-kilometer object released more than 10^8 megatons of energy and excavated a crater (Chicxulub in Mexico) approximately 200 kilometers in diameter. Among the environmental consequences were devastating wildfires and dramatic short-term perturbation in climate produced by some 10^{13} tons of fine dust injected into the stratosphere. However, we know from the fossil record that major mass extinctions occur at intervals of many millions of years. The chances of such an event taking place within, say, the next century are extremely low.

However, even projectiles substantially smaller than 15 kilometers across can perturb the global climate by injecting dust into the stratosphere, producing climate changes sufficient to reduce crop yields and precipitate mass starvation (but not a mass extinction). On the basis of the work of Brian Toon and Kevin Zahnle (NASA Ames Research Center), we can estimate that an impact by an asteroid or comet with an energy of a million megatons (diameter of about 2 kilometers) would produce a global calamity that might kill more than a billion people. Using this value and the known impact rate (from Figure 7.2), Clark Chapman (Planetary Science Institute) and I have calculated that there is about one chance in 4,000 that such a globally catastrophic impact

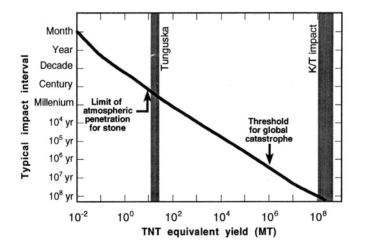

FIGURE 7.2

The same impact-frequency curve as Figure 7.1 with the addition of the 15-megaton Tunguska impact of 1908 and the >100 million megaton K/T impact of 65 million years ago. Also shown are the calculated limit of penetration into the troposphere for a stony (asteroidal) object (from C. Chyba and K. Zahnle) and the estimated threshold for a global environmental catastrophe, as defined in the text. For this value of the threshold, such a global catastrophe occurs roughly once per million years on average.

will take place in the next century and that, for an average individual, the chances of dying as a result of an impact are about one in 20,000. Phrased in terms of annual risk of death, this amounts to a little less than one chance in a million per year, or about the same as the chances of death in one round-trip commercial airline flight (Figure 7.2).

We have found that the total impact risk is dominated by objects a few kilometers in diameter, near the threshold for global agricultural collapse; smaller objects pose less risk, even though there are many more of them. The total impact hazard approaches that associated with other natural disasters, such as earthquakes or severe storms, suggesting that it is great enough to inspire public (and governmental) concern. Further, there is the qualitative difference between the globally catastrophic impact and all other natural dangers, in that only the impact has the potential to kill billions and destabilize civilization.

Although there is dispute among the experts about the accuracy of the risk numbers given here, the general trends are clear. Stepping back from the detailed calculations, we can state the following summary conclusions with considerable confidence:

1. Cosmic impacts represent an extreme (and relatively unfamiliar) example of the class of hazards with low probability but high consequences.
2. The statistical risk from impacts is substantially larger than the one-in-a-million lifetime risk of death often used as a threshold for government or regulatory interest.
3. Unlike any other known natural hazards, impacts can kill billions of people and endanger the survival of civilization.
4. The total risk increases with the size (energy) of the projectile; thus, any effort at hazard reduction naturally focuses on the very rare events associated with the larger impacting bodies.

Risk Reduction and Mitigation

Human reaction to the risk estimates given varies greatly, especially because the impact hazard represents such an extreme combination of low probability together with high consequence. Because no one has been killed by an impact in all of recorded history, it is easy to dismiss the risk as negligible and to regard those who express concern as alarmist. Further, the calculated annual risk of about one in a million is near the level at which many persons consider risks to be effectively zero. On the other hand, modern industrial societies spend large sums to protect people from even less likely hazards, ranging from hurricanes to terrorist attacks to trace quantities of carcinogenic toxins in food and water.

For most natural hazards, risk reduction or mitigation strategies can deal only with the consequences of the disaster. Thus, for example, we cannot stop an earthquake or even reduce its force, but we can mandate higher standards in building construction and develop plans to treat casualties and restore public services after the event. If impacts could be predicted weeks or months in advance, similar approaches could be taken, including evacuation of the populace from the target area. In addition, however, the possibility exists of avoiding the impact entirely by deflecting or destroying the projectile before it hits.

We discuss the probabilities of a large impact, but in reality this is not a Las Vegas game of chance. Either there is an object out there headed for the Earth or there is not. Any approach to this problem must therefore first consider the search for potentially hazardous asteroids and comets. Plans to augment current survey efforts have been presented, but funding is slow. As a result, only a handful of astronomers is actively engaged in the search for potentially catastrophic asteroids or comets. In fact, the total workforce devoted to this task on the entire planet is smaller than the staff of one McDonald's restaurant. Given that the survival of our civilization (including McDonald's) is at stake, our priorities should perhaps be reconsidered.

A survey for threatening objects is justified only if we can do something to avert a collision if one is predicted. Given a warning time of several years, it appears to be within our current technology to do so, by either deflecting the object or destroying it.

The most straightforward approach to deflection for an object in a short-period orbit is to apply an impulse that changes the orbital period. If such an impulse is applied several years before the threatened collision, only a very small velocity change (a few centimeters/second) is required. On the basis of discussions at several workshops that have examined this problem, the optimum way to impart such an impulse without risking accidental disruption of the body would appear to be a stand-off neutron-bomb explosion. Bombs of the appropriate yield exist within current nuclear arsenals, and in many examples that have been studied where warning time is ample, the required yields are quite modest (less than a megaton).

Before attempting to deflect an asteroid, we would almost certainly wish to study it in detail with scientific flyby or orbiter spacecraft. It would be especially useful to characterize its size, shape, and spin as accurately as possible before attempting the deflection impulse. Given sufficient warning, such prior investigation should be possible, and indeed the deflection itself could be undertaken in stages with multiple impulses applied.

The alternative approach of destroying a projectile requires the application of much larger energy. To avoid making the situation worse by converting the incoming object from a cannon ball into a cluster bomb, it is necessary to do more than simply disrupt it. Sufficient energy must be applied either to pulverize it (ensuring that no fragment is large enough to survive atmospheric entry) or to disperse all of the fragments so that none strikes the Earth. Destruction is the strategy of choice only when the warning time is so short that deflection is impractical. This is more likely to be the case for a cometary impactor than an asteroid.

Proposals to develop defensive systems against asteroids and comets raise a variety of issues, both philosophical and political. At the most basic level, we must decide if we wish to interfere with a natural process that has been important to evolution and that, in the form of the impact 65 million years ago, was essential to our own existence. Most people would agree that efforts at self-protection and self-defense are justifiable, and we take this position here. But what kind of defense system is appropriate to such a low-probability hazard?

In the case of asteroids and short-period comets, which dominate the hazard, it is clear that a low-cost survey is the first step. Only in the unlikely case that the survey finds a projectile on a collision course is further action required. A survey such as the Spaceguard Program proposed by NASA is a form of cost-effective insurance that protects our civilization against most cosmic threats. But not all. What should we do about the risk of a comet, descending from great distances and aimed to strike the Earth with a warning of only a year or two? How much should we spend for the extra insurance rider to cover this additional contingency?

There is no clear answer to this question, which is the focus of much of the current policy debate. There are those, among them Edward Teller (Stanford Research Institute), who advocate the immediate development and testing of nuclear deflection technology, leading toward the deployment of a planetary defense system early in the next century. Two classes of arguments have been raised in opposition: (1) arguments (with which I associate myself) based on cost effectiveness, questioning whether we can afford to spend billions of dollars on a defensive system that is unlikely to be used; and (2) arguments (with which Carl Sagan is associated) based on estimates that such a defensive system poses risks from accident or misuse that are greater than the impact danger it is designed to mitigate. This is an ongoing debate that is likely to intensify as more individuals and constituencies are drawn into it. For example, environmental activists have not yet joined this discussion, and it is unclear whether they will give more weight to protecting the planet from the ultimate environmental catastrophe of a large impact or protecting us from the incremental but more immediate risks of nuclear accidents associated with development and deployment of a defense system.

Life on a Target Planet

While we debate how to protect ourselves from future impacts, the Earth spins on its way through a minefield of potential catastrophes. Impacts have been important for the Earth throughout its history. The

Earth must have formed in the swirling debris of the solar nebula through a process of accretion of smaller bodies. High-velocity impacts generated heat, and eventually the upper layers of the planet melted to form a global ocean of liquid rock. At some point during the period of accretion, we were struck by another coalescing world about the size of Mars today – that is, with a mass about 10% that of the Earth. The smaller, Mars-sized planet was completely destroyed, and even the larger Earth was shattered to its core. Some of that ejected material continued to orbit the Earth as a giant ring, which cooled and collapsed to form our Moon.

If the Moon-forming impact had been just a little larger, the Earth itself would have been disrupted. There may have been examples of such planetary collisions during the early days of Solar System history, but if so, the direct evidence is long gone. We do see other examples of planetary peculiarities, however, that are best understood as the product of random collisions. Venus spins in the opposite direction from its orbital motion about the Sun, probably as the result of a late collision that struck it a glancing blow and reversed its direction of rotation, and the small planet Mercury appears to be the metal-rich remnant of a larger parent, stripped of most of its rocky mantle in another giant collision. It is largely a matter of luck that the final product of this chaos was the four inner planets we have today: Mercury, Venus, Earth, and Mars, plus the Moon.

The building blocks of life came to Earth through impacts, as studied by Chris Chyba (at the time at Cornell University) and his colleagues. In the outer parts of the solar nebula, far from the Sun, temperatures were much lower, leading to abundant water ice and other frozen gases such as methane, ammonia, carbon dioxide, carbon monoxide, and even ethyl alcohol. The volatiles on Earth must be derived from this distant reservoir in the outer Solar System through cometary bombardment in the last stages of accretion, which may have extended to several hundred million years after the birth of our planet. Most of the material of the biosphere – and of our own bodies – is comet-stuff, a gift from the outer Solar System. Were it not for this rain of ice and carbon compounds, our planet would be as dry and lifeless as the Moon. Life is a gift from the comets. But the gift did not come without a price to pay.

As the rain of cometary materials persisted, the Earth (and presumably Mars and Venus as well) built up a thick atmosphere of carbon dioxide and other compounds and developed shallow oceans of liquid water, rich in dissolved organic materials. Such an environment is exactly what most scientists think was required for the origin of life. The first self-replicating molecules must have formed in these early seas, perhaps on all three planets. If all of the impacting comets were small, this environment might have approximated the "warm little pond" hypothesized by Charles Darwin for the origin of life. However, the evidence preserved in the densely packed craters of the lunar high-lands suggests otherwise. At least a few of the impactors from that first half-billion years of solar system history were hundreds of kilometers across – the size of the largest asteroids and comets of today. Kevin Zahnle (NASA Ames) and his colleagues realized a few years ago that such large impacts were capable of drastically altering the terrestrial environment.

What happened when a 200-kilometer asteroid or comet smashed into the early Earth? Zahnle calculated that the energy of such an impact would melt and vaporize so much of the crust near the point of impact and that the planet would acquire a temporary atmosphere of rock vapor at a temperature of about 1,000°C. Under this terrible red-hot blanket, the oceans would be completely vaporized, killing any life forms that might have arisen. In effect, such impacts sterilized the planet. After a few hundred years the hot rock would cool and the oceans recondense, but the clock would be reset to zero for the origin of life. In an obvious understatement, this is called the impact frustration of life.

Extrapolating from the lunar record, it appears that the Earth was struck by a handful of such sterilizing impacts during its first half-billion years. Venus probably took as many hits, while Mars, being smaller, may have escaped such catastrophe. The last such impact probably took place about 3.9 billion years ago. It is extraordinary – and perhaps not coincidental – that the earliest chemical evidence of life on Earth dates from not too long afterward, at about 3.8 billion years ago. It appears that our kind of life formed not long after the end of this period of frustration. It is not too great an extrapolation from this evidence to suppose that life had formed several times previously, only to be wiped out by a sterilizing impact.

Subsequent to the end of the heavy bombardment, the Earth has continued to experience occasional impacts. Some of these, like the Cretaceous impact of 65 million years ago, severely disrupted the environment and redirected the course of biological evolution. Impacts, taking place at the right level, appear to have played a crucial role in the history of life. Too many impacts, and the planet is sterilized. Too few, and evolution may stagnate in a static, benign environment. Just the right number of impacts, and we have the development of *Homo sapiens*.

We are thus led to a remarkable picture of the role of impacts in the history of the Earth, a role only recognized during the past decade. Our planet was formed by the accretion of impacting debris; the Moon was blasted from the Earth's surface in a giant impact not long after the planet formed; the rain of comets from the outer Solar System subsequently carried life-giving water and organic compounds to the inner Solar System, but at the same time subjected the Earth to a terrible bombardment of projectiles, the largest of which boiled away the oceans and sterilized the surface; and continuing impacts by objects no more than tens of kilometers across have caused mass extinctions and created opportunities for evolutionary change and diversification. The current debate about protecting Earth from impacts is only the latest chapter in a long and violent history of the planet.

8
Extraterrestrial Intelligence: The Significance of the Search

FRANK D. DRAKE

University of California at Santa Cruz

My subject is the significance of the Search for Extraterrestrial Intelligence (SETI); I will also explain how we're actually going about the search. To most of us, the significance of the detection of another extraterrestrial civilization is in fact well known. The significance of such a contact depends markedly on just what kind of a person you are and what your interests are. If you're a member of the general public, perhaps the most appealing thing about it is the adventure that would occur, the adventure of contact with another species with an entirely different history and a different physiognomy, living on a different planet, with different technology, different forms of government. What an adventure! All through human history we have been excited when different human cultures have come in contact with another. A SETI success would produce the same kind of event, but on a level far higher and more exciting than anything we have experienced before. This is, of course, the grist of many science fiction movies and books.

On a more philosophical note, such a contact would go far toward answering the question: What is the place of human beings in the universe? To what degree are we average, normal? What potential do we have? What is our destiny? What could we achieve? This is something we can learn from contacting and studying other civilizations, particularly those that are more advanced than ourselves. In the same way, such a contact will open our eyes to possible futures for ourselves as well as hazards of the kind that David Morrison described. What becomes of intelligent civilizations? Do they colonize space? Do they terraform their entire planetary system to provide living room for their inhabitants? In what ways do they act so as to achieve, we expect, the highest quality of life for their creatures?

Possible futures: That is one of the great significant impacts of an eventual detection. But to scientists, of course, of very great significance is that such a contact would open to us the great library of knowledge that must exist among the intelligent civilizations of our galaxy and other galaxies – information of a scientific and technical nature, as well as philosophical and biological, gathered not only over hundreds of years, as has been our experience, but over literally thousands, millions, even (I won't use the "B" word), thousands of millions of years.

Surely from this would come a richness of knowledge, even wisdom, that would help us achieve a higher quality of life, save our resources, give us sophisticated information we would like to have, and within years rather than hundreds of years, or thousands, which it might take us to achieve these same results through our own research.

So those, in a nutshell, are the possible significant results of SETI, and of course they are of the most profound consequence and importance to all of us. That is why those of us who have the privilege to work in SETI feel it is a very important project and we give ourselves to it.

Now, these ideas are not new. They have been around a long time, as long as people have first understood, as Galileo did, that the other planets were in some way like the Earth. The ideas became even more widespread when we recognized that the stars were objects like our Sun. As the years have gone on, people have speculated or fantasized about life in the universe and particularly intelligent life. An example of early excitement about this subject is the front page from *The New York Times*, August 17, 1924. (Walter Sullivan is not responsible for this.) It contains drawings, totally wrong, done by Percival Lowell and this very chauvinistic statement that Mars invites mankind to reveal "his" secret. The following Saturday, there was to be a very close approach of Mars to the Earth, and all the world looked. Among all these observations were many attempts to detect signs of intelligent life on Mars. We read, for example, that from the top of the Jungfrau, Swiss astronomers were to flash a light signal to the planet with a giant helioscope.

At the same time, there were many attempts to detect radio signals from Mars. Searches for extraterrestrial radio signals are not something solely of our era; they started actually back in the 1890s and reached a peak in 1914. On another front page of *The New York Times*, from seven days later, there is this announcement, front page news, "Radio Hears Things as Mars Nears Us." We're supposed to be impressed, because it tells us that it took a twenty-four tube set in England to pick up the strong signals and that Vancouver, for some reason, was of a special interest to the Martians! We don't know why that is, and, in fact, the great cryptographer, William Friedman, who had broken the German codes in World War I, was standing by in Washington to decode the messages. Front page news, but as is typical, when it turned out those dots and dashes were from a radio transmitter in Seattle, that was quietly announced on page thirty, where one could read: "Mars

Sails By Us without a Word. Disappointment." These events produced one of the false alarms that discredited not just searching for life in the universe, but planetary science as a whole for many, many decades. That science only recovered its prestige and escaped its taboo nature with the dawn of the Space Age.

It is interesting to point out that at this very same time, something of very great importance to SETI was occurring. Again, in a front page of *The New York Times*, with the headline: "Talk with Aviator by Radio a Mile over Central Park," and "Experimenter Catches Words above the Roar of the Airplane." This may not seem very startling today, but if one looks at the engineering textbooks of that time, one finds that the physics of radio links was very fuzzy. The engineers believed that the design of an electromagnetic communication link required a transmitter and a receiver, each of which had a connection to the ground. There was this false concept that somehow radio signals were conducted in a sort of electrical circuit in which one leg of the circuit was in the Earth and the other was through the atmosphere; therefore, without that ground connection one could not communicate. Of course if that were true, it would have terrible consequences for SETI, because there would be no way one star could communicate with another. But we read that the flyer did broadcast directly from the plane. When you read this you find that there was a Lieutenant Connell, on his hands and knees in the open cockpit of this airplane, shouting into a microphone so he could be heard over the roar of the Liberty engine. It was a great moment for SETI. He succeeded with no ground wires! That was a very important breakthrough because it showed that radio didn't need ground wires and so radio could communicate between the stars. The technological basis for SETI was established.

We sense again the interest in the possibility of electromagnetic communication between the stars from a front page of *The New York Times*, 1933, announcing the discovery of cosmic radio emission. Surprisingly, reporters and editors recognized the significance of Karl Jansky's discovery of cosmic radio emission and put it on the front page. The headline reads, "New Radio Waves Traced to Center of the Milky Way, Mysterious Static Recorded by K. G. Jansky Held to Differ from Cosmic Ray." The lowest headline says, "Only Delicate Receiver is Able to Register; No Evidence of Interstellar Signaling." Even then, they wondered if perhaps radio was the means by which we might contact other civilizations.

All of these early ideas were very speculative and they were not properly scientific in that there were no quantitative calculations made whatsoever of, say, what power levels were required to make a detectable signal over what distance. These were really hand-waving ideas. It was only in the 1950s and early 1960s when people started to approach the matter in a proper scientific way, in the manner first done in print by Philip Morrison and Guiseppe Cocconi at Cornell.

Another watershed event at that time was the publication of the book by Carl Sagan and Iosef Shklovskii on SETI, a monumental book that has become a classic. Here Carl took a rather narrowly written book by Shklovskii, written in the context of the limited state of knowledge of the Soviet Union, and added to it a great breadth of scientific and technical knowledge. This created what, to this day, is a handbook for the quantitative analysis of what it might take to detect life elsewhere in the Solar System. So, in the late 1950s, early 1960s, finally, this subject became one that had a proper quantitative scientific basis. We entered what might be called the modern era of SETI.

Now, along the way, many discoveries have stimulated interest in this subject and have indicated or provided growing evidence that intelligent life can be expected to be ubiquitous in the Milky Way and in other galaxies. We've learned that the galaxy is fifteen billion years old. That there are four hundred billion stars, many like the Sun. We have learned how star formation proceeds and that in the clouds of gas and dust that form the stars, we have the chemical precursors of life on Earth. There have been discussions in the previous chapters about these materials being brought to Earth by the comets. We've seen evidence of other solar systems, although not yet another solar system just like our own. A great deal of circumstantial evidence suggests that planetary systems and planets like the Earth are extremely common. We've seen in the planets of the Solar System themselves a great abundance of the organic molecules and other materials likely to give rise to life, and even to intelligent life. We know about tantalizing Titan, motivating us, suggesting that the abodes of life are very abundant in the Milky Way and therefore that searches for life are of great value.

SETI has significance in areas that are less profound than the ones I first mentioned. It has provided grist for the entertainment industry. The best-selling movies of all time have been based on ideas of SETI. Even now, Carl is working on such a movie, *Contact*, which will be a source of entertainment, but will also build understanding in the population at large about the possibilities of contact with extraterrestrial life. That helps us in carrying out our search. SETI has also turned out to be a wonderful magnet to attract young people to an interest in science. In many places, including our SETI Institute, curriculum materials based on SETI are being developed. These turn out to work extremely well with young people in the real world, because young people are very interested in life in the universe. Once they find out that to understand SETI you have to learn about organic chemistry, or atmospheric structures, and so forth, they're willing to study those things that might otherwise be boring.

The existence of SETI activity has stimulated much theoretical work about the propagation of radio waves in the galaxy and more speculative theoretical work, some of it good, some of it not so good, which has to do with the nature and behavior of life in the universe.

SETI is a source of pride in cultures and countries, because we recognize that the pursuit of extraterrestrial intelligent life is one of the most noble, if we can use that word, of human tasks. Both America and the Soviet Union have taken great pride in it, and so have states like West Virginia. A pamphlet for potential tourists to that state points out "buried treasures." It describes the National Radio Astronomy Observatory as a place where "they listen for signs of life in outer space." You might think they don't do anything else there, according to this pamphlet! They just listen for signs of life in outer space! Actually, no major program of this kind has been carried out there in the last twenty years!

Other impacts of significance: SETI brings out the very best in people and attracts some of the most talented of our scientists and engineers. Among these is Dr. Jill Tarter, who is the senior project scientist for Project Phoenix, the new name for what was the NASA SETI Search. That search, having lost its federal funding, has become a privately based project operated by the SETI Institute. Jill Tarter's original name was Jill Cornell, and in fact she is a direct descendant of the founder of Cornell University, Ezra Cornell, and got her education in engineering physics at Cornell. Dr. Kent Cullers, Ph.D. in physics, is in charge of the signal detection systems for Project Phoenix, a very daunting task. In Project Phoenix there are some fifty-six million channels of information providing data. Kent Cullers' task is to find algorithms that can search through this in real time for signs of intelligent signals, yet in an affordable way. He has been blind since birth, yet he carries out the most abstruse, difficult mathematics in his head, all the mathematics connected with communication theory and computer algorithms. Dale Corson may be surprised to be mentioned in connection with SETI, but he has been important through his role in the construction and the upgrading of the Arecibo telescope reflector. He helped make what is to this day and for the next few decades probably by far the largest radio telescope in the world. It is the preeminent telescope for SETI, the one that gives us our greatest power, the one that we wish to use the most. Project Phoenix has contributed about two million dollars to the further upgrading of Arecibo now under way.

Dale Corson and the other people at Cornell have helped in other ways. For example, Jill Cornell wanted to study engineering physics at Cornell. In the will of Ezra Cornell it was specified that all male descendants of Ezra Cornell shall be admitted to Cornell and will attend free of all tuition. Jill inquired into this and its lack of political correctness, and of course there was a great deal of discussion in the back room about what to do. In the end, Dale Corson, as president, recognized the unfairness of that stipulation in the will, and quietly took care of it all. As a result, Jill did attend Cornell on a full scholarship and that was an important turning point for SETI, because without Jill, SETI would be a much weaker endeavor.

As has become obvious, the name Cornell pops up repeatedly. Many of the people active in SETI had their roots there: Phil Morrison, Carl Sagan, Jill Tarter, myself, and others. This says, if you think about it, that it's not really a result of chance but the result of the fact that at Cornell University creativity and original ideas have been nurtured and protected. Cornell has protected and nourished the pioneers, the people who have interesting but unusual ideas, and from that has come much of the university's greatness, not just in SETI, but in many other areas.

What is another significance of SETI? SETI works as a motivator for the development of the very best in frontier technology, particularly in radio receiving systems. The first modern SETI system, used in 1960 in Project Ozma, was rather simple. Even so, it cost $2,000 and occupied four banks of equipment, all using vacuum tubes, by the way. See Figure 8.1. This was before the transistor was invented. It could monitor one channel at a time. The people who did this experiment are shown in Figure 8.2 at a reunion of a few years ago in front of the 85-foot telescope where it took place. In the thirty-four years since Project Ozma, other activities have joined SETI in motivating the development of much better technology such as the Arecibo telescope, which has one hundred times the collecting area of that of the Ozma telescope. A picture of its great reflector is shown in Figure 8.3,

FIGURE 8.1

Radio receivers used during the 1960 Ozma Project.

FIGURE 8.2

The Tatel 85-foot radio telescope in Green Bank, West Virginia, and the team that was involved with the Ozma Project in 1960 (the author of this article is second from the right standing).

the reflector reflecting rays to the suspended platform, which is right now undergoing major alterations.

With the construction of such telescopes SETI has been given the ability to detect signals from anywhere in the galaxy, signals no stronger than we radiate. Many years ago it was recognized that the real challenge to SETI was not sensitivity so much, but the ability to

FIGURE 8.3
The Arecibo radio/radar telescope with a diameter of 1,000 feet. Located in Puerto Rico, this giant telescope is operated by Cornell University for the National Science Foundation.

search through all the possible frequency channels in the microwave spectrum where signals might occur. That has led to the application of some of the most brilliant minds on the planet, and an article is presented by one in the next chapter, to the development of systems, affordable systems, which can achieve that task. A picture of one of the very first is in Figure 8.4, using digital circuitry in this case to develop a system that monitors about 68,000 channels at once. This system was produced by Paul Horowitz, using mostly very simple off-the-shelf components. Over the years, it has led to much more powerful and complicated systems. These require special-purpose computer chips that have been developed at the cost of about $300,000 for the first one. After that, they cost $50 each, and they are about the size of a postage stamp. They do eighty million flops and 220 million data transfers a second, all to simulate a multichannel radio receiver. One such chip has the same power as a Cray 1 supercomputer programmed to do the same task. We have now incorporated these in our systems, forty-seven of them, so we have in effect forty-seven Cray supercomputers to produce the instruments of Project Phoenix. The equipment again occupies four racks, just like the Ozma receiver did, but in this case the total number of channels is fifty-six million. See Figure 8.5. This system is 100 trillion times more powerful than what we had thirty-four years ago. In fact, the doubling time for improvements has been consistently 250 days; exponential growth has continued. The Project Phoenix equipment is in a trailer that can be carried in a large transport plane. See Figure 8.6. Right now it is being readied to go to Australia for six months of observations with the Parkes Telescope, the last chance

FIGURE 8.4
A digital system that can monitor 68,000 spectral channels at once.

FIGURE 8.5
The fifty-six-
million channel
receiver used in
Project Phoenix.

FIGURE 8.6
The NASA trailer
containing the
Project Phoenix
complex
instrumentation
ready to be shipped
to the Parkes
radiotelescope in
Australia.

for many years to observe the southern candidate stars where we might find an extraterrestrial intelligence signal.

SETI people are motivated to dream of much bigger systems. We dream of great systems on Earth, as may be built some day as the significance of SETI becomes more widely recognized. Perhaps there will be SETI in space; systems have been in the eyes of the designers for many years now. The ultimate dream is that we will achieve those very profound results I mentioned earlier.

Lastly, there is one other significant aspect of SETI that Carl has emphasized many times, which comes about when we do SETI and construct messages such as the Pioneer 10 plaque. Carl and I invented that one day during a coffee break in a corridor at a American Astronomical Society meeting (I probably shouldn't reveal that; I probably should say there were endless committee meetings and white papers and planning documents, but that's not so; it all happened over coffee in a hotel). When we not only search but send messages to space, we are actually sending messages to ourselves, reminding us of what human beings can achieve if they work together and use their talents in the very best possible way. SETI sends a message to us that humans can achieve great things. If we work hard enough we can join the company of other creatures in space and receive all of the great bounty that will accrue when, finally, the discovery is made.

9

Extraterrestrial Intelligence: The Search Programs

PAUL HOROWITZ
Harvard University

The topic, "Extraterrestrial Intelligence: The Search Programs," should also include the searchers – you can't really disembody the personalities who motivated these searches and the people who did them from the searches themselves. I'd like to highlight some of the giants of SETI, and their predecessors – Hertz, Jansky, Purcell – names you don't hear so much nowadays. That means many pictures of people, their antennas, and their equipment, and only a few equations or graphs.

In fact, let's get the equations over with; Figure 9.1 illustrates SETI Fact Number One: SETI is possible because, as each of us who have become delighted with the subject has discovered rather early on, radio communication is extraordinarily efficient. All of the equations crammed into one figure. It's an old calculation that those of us in this business have all done, in one form or another: You take a pair of modest-sized radio telescopes, a couple of hundred meters in diameter (that's smaller than the Arecibo dish, though not by a lot); you space these apart a modest distance, let's say a thousand light years; you transmit a three-centimeter wave, say; and you ask, for a given amount of energy transmitted, how much is received? Is this crazy scheme going to work?

You calculate transmitted energy, areas, gains, and all that sort of thing. And when you're done, you find that a dollar's worth of energy transmitted results in 4×10^{-12} ergs of energy received out there. Now you should ask yourself, "Well, ten to the minus 12 ergs, not very many ergs, and ergs are pretty darned small; is this even detectable?" Well, you'd be right, it's not a lot of energy; but the crucial point is that the energy *that it has to compete with* is even less. The received signal has to compete with cosmic noise, antenna noise, amplifier noise, and so on. How large are these? The delightful fact is that, at these microwave frequencies, you can build receiving systems with large apertures (Arecibo-sized) whose equivalent thermal noise temperature

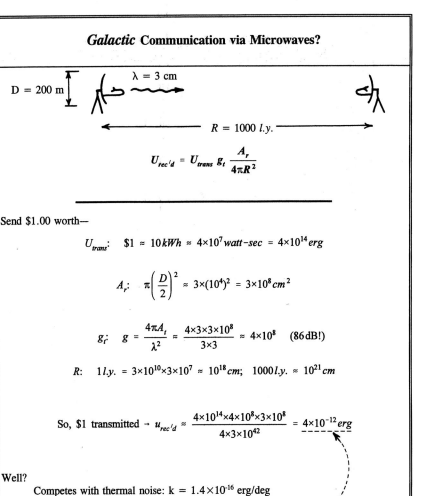

FIGURE 9.1
A simple calculation demonstrates that communication by microwaves is extraordinarily efficient.

is of the order of a hundred Kelvin or less. And that includes the combined noise contributions of the dish, feeds, amplifier – the works. It is an astronomical fact that the sky itself is extremely cold at these frequencies (which, of course, motivates this choice of wavelength).

So, finally, the thermal fluctuation energy that this message has to compete with is something like 10^{-14} ergs, four hundred times smaller than the message itself. In other words, a dollar's worth of energy is

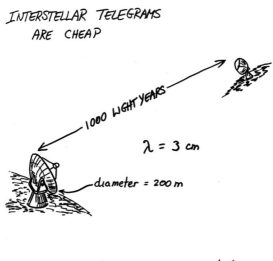

INTERSTELLAR TELEGRAMS
ARE CHEAP

$\lambda = 3$ cm

diameter = 200 m

ENERGY RADIATED PER BIT TRANSMITTED : 0.4 kWh

COST PER WORD : \$1.00

FIGURE 9.2
Even with
technology no
better than we have
now, galactic
communication is
a bargain (after
E. M. Purcell).

all it takes. For a dollar you've got four hundred times the fluctuation energy at the receiving end. You pick your bit rate and coding so that each bit is received at fifteen times the noise, say, and that gives you something like twenty-five bits per buck received at the far end. If you group these into letters of five bits each (don't use ASCII; use Baudot!), and if you speak in words of one syllable, you wind up with the conclusion that interstellar telegrams are cheap (Figure 9.2, an old slide of Ed Purcell's). This is a remarkable, and essential, fact of SETI – that an interstellar telegram out to a thousand light years costs a mere dollar per word, and that's using only technology that we have on earth now.

In fact, we have even larger dishes. The largest on earth is the great 1,000-foot Arecibo dish, with double the area of those I have been discussing. With transmitters and receivers no better than we have right now, two of these things could communicate anywhere across the galaxy (assuming, of course, that they knew to point at each other and to use the same wavelength). This fact is, for most people, counterintuitive.

Fortunately for SETI, our galaxy is endowed with an ample supply of candidate life sites; it is shown (Figure 9.3), somewhat simplified. It contains something like four hundred thousand million stars, in various colors. It's a flattened disk a hundred thousand light years in diameter; we're out in the galactic suburbs, so to speak, at about thirty thousand light years. The Sun is in the plane, as shown, in the middle of a more or less uniform distribution of stars; there are roughly a million stars like the Sun within a thousand light years. That is the range within which we can communicate for a dollar per word.

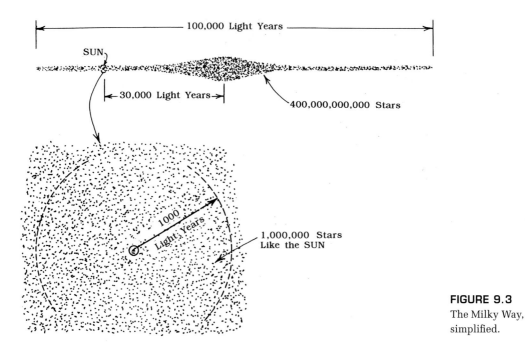

|← ———————— 100,000 Light Years ————————— →|

SUN

|← 30,000 Light Years →|

400,000,000,000 Stars

1000 Light Years

1,000,000 Stars
Like the SUN

FIGURE 9.3
The Milky Way,
simplified.

We optimists feel that SETI is likely to be successful, because we think that the odds are very good that other advanced life exists elsewhere in the galaxy. Many of us expect the nearest such civilization to be within that thousand-light-year neighborhood. It's traditional to talk about this problem in terms of the celebrated Drake equation, with its estimates of habitable planetary systems, origins of life, evolution of technology, and so on.

I'd like to sidestep all that and show, instead, Figure 9.4, drawn by my ever-creative colleague Purcell. He plotted the world lines for a simple model of galactic demography, in which one sunlike star in 10,000 has radio technology for 10,000 years, some time during the past ten billion years. Time increases upward, with the upper left corner being here and now. Part of the lines represent life-bearing planets. These parts have not yet developed technology. At some point they turn "radio active" and for 10,000 years they transmit and then they stop. Don't ask why – maybe they've all become philosophers or something. Now the interesting question – are any of those on our backward-looking light cone? In other words, can we hear from any of them? As you can see, even with this somewhat pessimistic scenario, we catch a half dozen transmitting civilizations in our net.

So, on this speculative issue of other advanced life many of us believe that it's likely, and even overwhelmingly likely, that there are other civilizations out there. And, as we demonstrated at the outset, it's not a matter of speculation at all, but of technological fact, that

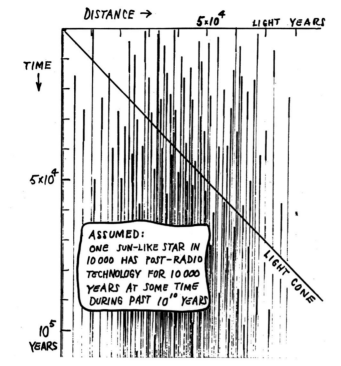

FIGURE 9.4
Galactic
demography,
Minkowski-style.

we've got everything we need right now to communicate with them, providing we guess correctly about how they're transmitting.

It's curious that the speculative aspects seem so to dominate the literature! Much (*too* much) has been said and written on "Drake-ology." It's fitting on this occasion to note that the landmark work on the subject is probably Shklovskii and Sagan's 1966 *Intelligent Life in the Universe*, an extraordinary book. There are three grand subject areas. The first is called "The Universe" – eleven chapters. Then "Life in the Universe" – eleven more chapters. And finally, "Intelligent Life in the Universe" – the last eleven chapters. This is your basic guide to everything you ever wanted to know. Thirty-three chapters, done 10,000 miles apart, without FedEx or e-mail.

Of course, these ideas had been around for a long time, centuries, even millennia, in rather squishy form. But they had only recently been addressed quantitatively, by Cocconi and Morrison in their famous 1959 article, and experimentally by Drake in his 1960 Ozma search. This brings us to The Search Programs.

Let me mention a few pioneers who are not normally resident in the SETI pantheon. Heinrich Hertz, through a series of ingenious and wonderful experiments between 1886 and 1892, set out to try to verify or disprove the then-controversial Maxwell theory. He generated electromagnetic waves. He measured the propagation speed and found it to be close to the speed of light, thus deducing that electromagnetic

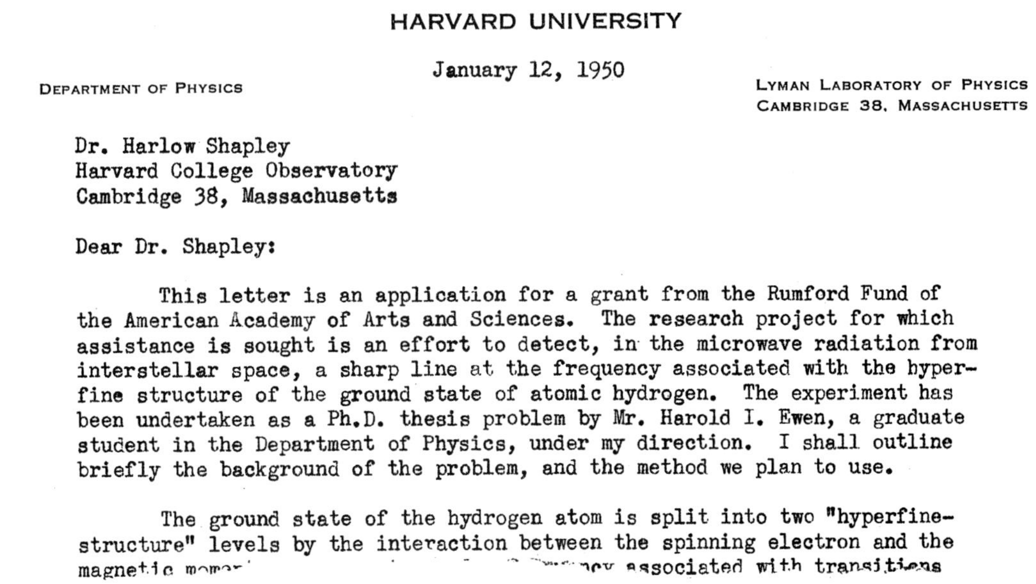

HARVARD UNIVERSITY

DEPARTMENT OF PHYSICS   January 12, 1950   LYMAN LABORATORY OF PHYSICS
CAMBRIDGE 38, MASSACHUSETTS

Dr. Harlow Shapley
Harvard College Observatory
Cambridge 38, Massachusetts

Dear Dr. Shapley:

        This letter is an application for a grant from the Rumford Fund of
the American Academy of Arts and Sciences. The research project for which
assistance is sought is an effort to detect, in the microwave radiation from
interstellar space, a sharp line at the frequency associated with the hyperfine structure of the ground state of atomic hydrogen. The experiment has
been undertaken as a Ph.D. thesis problem by Mr. Harold I. Ewen, a graduate
student in the Department of Physics, under my direction. I shall outline
briefly the background of the problem, and the method we plan to use.

        The ground state of the hydrogen atom is split into two "hyperfine-
structure" levels by the interaction between the spinning electron and the
magnetic mo~~~~~   ~~~~ ~~~~~ associated with transitions

FIGURE 9.5

A bit of history: Purcell's grant application for the discovery of galactic 21-cm radiation.

radiation and light were probably one and the same thing. He measured refraction, reflection, polarization, standing waves – all with
nineteenth century apparatus. Karl Jansky, in the early 1930s, became
the world's first radio astronomer; he had a radio telescope. This really
marks the founding of SETI, radio waves detected from outer space.
All we need is someone to send them to us.

    I discovered a wonderful little piece of history very recently. I ex-
perienced some of the thrill that historians must feel when they open
a drawer and find some yellowed pages telling a story no one has seen
in fifty or a hundred years. I did a little bit of snooping (with permis-
sion) in Ed Purcell's file drawers the other day, and I found the most
wonderful set of letters and pictures. The letter in Figure 9.5 was written in 1950, a date radio astronomers will recognize as just prior to the
discovery of the 21-centimeter line. Ed is applying to Harlow Shapley, head of the Rumford Fund at the Harvard Observatory. It's a grant
application! He gets it out right in the second sentence. "The research
project for which assistance is sought is an effort to detect, in the microwave radiation from interstellar space, a sharp line at the frequency
associated with the hyperfine structure of the ground state of atomic
hydrogen." He says this is going to be a Ph.D. thesis of Mr. Harold
Ewen. Further down, he says, in a wonderful Purcellian manner, "Microwave radiation at this wavelength can be absorbed or emitted by
free neutral hydrogen atoms, of which interstellar space contains a
supply abundant for our purpose." People don't write grant proposals
like that any more.

    Well, he goes on, again just a few highlights here. "I have computed
the transition probability and on the basis of available astrophysical

evidence, I believe there is a good chance that the line can be observed." Of course we all know, now in retrospect, this became the bread and butter of radio astronomy. "The antenna itself will consist of an electromagnetic horn mounted outside the upper floor of the Lyman Laboratory." And he says, just in case Shapley needs any convincing, "I need not point out to you the astrophysical implications of the experiment if successful." Need not, perhaps, but he did anyway. Now he worries a little bit about the competition – "An experiment of the sort described has been in the minds of many people, I am sure, and it is not unlikely that someone will beat us to it" – simultaneously displaying his fondness for the double negative. "However, we have set ourselves the limited objective of detecting the line, if possible. Clearly, once the existence of the effect is established, many more elaborate investigations would suggest themselves." *That's* understatement. And finally, he gets to the punch line – "... and it is to defray these expenses that I request a grant in the amount of $500." He says at the end, "I hope that the Rumford Committee will feel that this project is a suitable one for support by the Rumford Fund, and is worthy of the assistance requested." In one of the greatest bargains in the history of radio astronomy, $500 is what it cost to get the 21-centimeter line!

Six weeks later we have the reply (Figure 9.6) from the Rumford Fund secretary (they really moved on grants in those days) in which he has approved the grant of $500. And the rest is history. I have some vintage photographs. The horn antenna in Figure 9.7 is outside the Lyman Laboratory, with graduate student "Doc" Ewen in charge. It's definitely an amateur job -2×4s, plywood, stuff hanging off. On the other side of the wall was this contraption (Figure 9.8), all vacuum tubes of course, complete with headphones, a Simpson VOM (for old timers), and other accouterments of 1950 communications technology (such as World War II radar jammers used as local oscillators).

And here is what they found (Figure 9.9). The detection of galactic hydrogen is the wiggly line that crosses the slant baseline, above and then below, in a classic differentiated signature caused by Dicke switching. The publication that followed in *Nature* is worth a few remarks. In the curious style of the time, the humble title "Observation of a line in the galactic radio spectrum" is followed by a subtitle "Radiation from galactic hydrogen at 1420 megahertz per second." You can imagine a sub-subtitle: "Scientists Amazed"! In the short report Ewen and Purcell say, after some preliminaries, "We can now report success in observing this line"; and finally, at the end of his article, with appropriate gratitude they thank the American Academy for the Rumford grant, the $500 that made the discovery possible.

This was submitted on June 14, but not published until September. Why the delay? And why, curiously, does the very next article in the same issue have the similar title "Neutral hydrogen hyperfine detection." What's going on here? Well, it turns out that the Dutch group,

AMERICAN ACADEMY OF ARTS AND SCIENCES
28 NEWBURY STREET
BOSTON

February 28, 1950

Dr. E. M. Purcell
Lyman Laboratory
Harvard University
Cambridge 38, Massachusetts

Dear Dr. Purcell:

I have the pleasure to inform you that the Council
of the Academy at its meeting on February 8, 1950, voted
to approve a grant of $500 from the Rumford Fund to you
to assist in your microwave experiments on radiation
from interstellar space.

Payment on this grant will be made to you in the
manner you indicate upon your application therefor to
the Treasurer of the American Academy of Arts and Sciences, 28 Newbury Street, Boston 16, Massachusetts.

Sincerely yours,

John W. M. Bunker

John W. M. Bunker
Secretary

JWMB:C

FIGURE 9.6
The American
Academy of
Arts and Sciences
makes a
historic grant.

knowing of the Harvard discovery, were able to replicate it six weeks
later. Purcell and Ewen then insisted that *Nature* delay the publication of their own paper until the other discoverers had a chance to
write up their results and have it published simultaneously. A most
gentlemanly gesture.

Well, it's not *his* sixtieth birthday, it's Carl's, so let me get on with
it. That was 1951. In 1959, the landmark Cocconi and Morrison paper
(or, as we like to say in Cambridge, the Morrison and Cocconi paper)
appeared, suggesting that efficient communication with extraterrestrial civilizations by means of microwaves is possible, for many of the
good reasons that we understand since then.

In 1960, Frank Drake made history with his one-channel, two-
star search, called Ozma. It was the first search of the modern era —

FIGURE 9.7
The first antenna on earth to detect radio waves from neutral hydrogen in space.

informed by radioscience and communication theory – the first search that could have succeeded. Among several legacies of Frank's experiment is the apparent mandate that SETI apparatus always occupy four racks. I'll demonstrate this curious effect in subsequent photographs.

This opened the floodgates (or perhaps I should say the tricklegates) for subsequent searches, most of which have been done at the hydrogen hyperfine frequency, or its near relatives. The traditional reason one always hears is that signaling is most efficient in the centimetric region of the spectrum (and it is). To be historically accurate, however, we should note that Frank chose hydrogen for a different reason: He wanted deniability if accused of wasting government money for crazy research – this was, after all, a *radioastronomy* receiver.

But this is in fact the efficient wavelength regime. Drake showed that there is no obvious advantage to short or long wavelengths, *on the basis of power delivered to a distant observer*. What matters, of course, is the competition – the galactic and atmospheric noise backgrounds, which clearly favor centimetric wavelengths if you believe that an efficient strategy is optimum.

Let's take now a quick tour through a rogue's gallery of searchers. Palmer and Zuckerman were persistent guys using the NRAO 300-foot radio telescope. They did 500 hours of searching, 674 stars, 21

FIGURE 9.8
State-of-the-art
radioastronomy
laboratory (1950).

centimeters, back in the 1960s. They were rather enthusiastic then, but
Ben, anyway, has become grizzled and wizened and rather pessimistic
since those heady days. He's now the grouch (though a friendly grouch)
of SETI; and as if to underscore the futility of it all, their antenna
even fell down. While the antenna was still up, Jill Tarter and Jeff
Cuzzi did a search there, using VLBI equipment to log the data. These
people looked for a hundred hours at 200 stars, which was a very
exhausting effort. If you're going to do long SETI observations, you
really shouldn't be sitting there; let your compulsive computers do
the boring stuff.

Here's the longest-running search of them all. This unusual view
(Figure 9.10) shows part of the Kraus antenna at Ohio State University.
A tiltable flat panel combines with a fixed curved antenna, forming a
focus at a movable feed horn on a railroad carriage, in order to track
as the Earth turns. This is a way of getting a very large aperture on the
cheap – you hire people who put up billboards, asking them to put
something shiny on it. These guys never quit – they've been searching
since 1973, 21-centimeter wavelength again, all sky, all the time. Jill
Tarter and colleagues have used the similar antenna at Nancay in a
series of searches.

Long-distance radio listening attracts amateurs, too. Bob Stevens
bought for one dollar a 60-foot early warning radar up in the Northwest

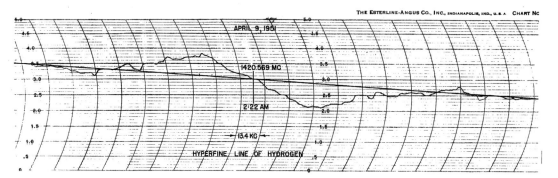

FIGURE 9.9

Galactic hydrogen drift scan, two weeks after initial discovery.

Territories; he hung in there, living in the shack next to it as long as he could, finally leaving town emaciated and frozen nearly solid. And Bob Gray has a little dish in his back yard in Chicago. Everybody can get into this act.

Now, there's one antenna conspicuously missing – Arecibo. An interesting search was done there, but you don't hear too much about it. In 1975, a certain Sagan and Drake published an article in *Scientific American* on "The Search For Extraterrestrial Intelligence," in which they say that "there can be little doubt that civilizations more advanced than the earth's exist elsewhere in the universe" and that we ought to go look for them. And they point out that it may make sense to look at whole galaxies because the strongest signals as observed from Earth may not be at the nearest stars: There may be incredibly radio-luminous civilizations in other galaxies. What they didn't say, perhaps out of shyness, is that in fact they were doing that very experiment themselves! Here is their very brief report, in its entirety, from an Arecibo Observatory Quarterly Report in 1975:

> "Search for Signals from Extraterrestrial Life" (F. D. Drake and C. Sagan)
>
> A brief run on the night of March 24/25 served as the first shakedown of the new autocorrelator and a test of detection sensitivity. Normal terrestrial communication signals reflecting from the moon near 430 MHz were picked up with the Arecibo telescope and analyzed with the first 252 channel quadrant of the correlator. Later in the evening various galactic and extragalactic sources were observed at 1420 MHz. No non-terrestrial narrow-band signals were detected.

Note particularly the first sentence – you always had to have an excuse for doing SETI in those days! I searched everywhere for a photograph of Carl with his hand on the switch. I tried the Arecibo archives, the Harvard archives, the archive's archives – there's no photograph of Carl with his hand on the switch. Maybe he never had his hand on the switch, except that night.

FIGURE 9.10
Tiltable flat
reflector of Kraus
antenna at Ohio
State University;
its collecting area
is equivalent to a
175-foot dish.

There have been roughly fifty searches to date; Table 9.1 gives the flavor. Starting with Ozma in 1960, Frank Drake's single-channel receiver (which was a helluva lot more sophisticated than presented in his talk. That wasn't fair at all! He used a Dicke switched scheme, with double horns, and switched wide/narrow band detectors, a lovely piece of work); through the Ohio State work starting in 1973, and so on, these are the radio wave searches, mostly in the region of neutral hydrogen. But there have been other searches – the world is not completely hung up on centimeter radio waves, even though they do look awfully good. People have looked for optical pulses, for lines from ultraviolet lasers, for gamma bursts, for tritium hyperfine radiation,

TABLE 9.1
MICROWAVE SETI AND OTHER SEARCHES

	21 cm (HI), etc.		Other	
1960 Ozma	1 channel, 2 stars		Radio pulses	
			Optical pulses	
			UV laser lines	
1973 Ohio State	50 channels,		γ bursts	
	all-sky		("skid marks")	
	~50 searches		Tritium hyperfine λ	
			1.5 GHz	
1985 META	8×10^6 channels,		Junk at libration points	
	all-sky		Optical	
1992 Serendip III	4×10^6 channels,		Radar	
	most sky			
····now····			Dyson spheres (IR excess) 8–14 μm	
1995? BETA	240×10^6 channels,		mm RF: H_2O, e^+e^-, etc.	115 GHz
	all-sky			203 GHz
1995? Serendip IV	130×10^6 channels,		Fission product spectral	
	most sky		lines: optical	
			IR lasers (CO_2) 10 μm	
1995? Phoenix	56×10^6 channels,			
	1000 stars,		$^3He^+$ spin flip 8.6 GHz	
	all modes			
	$\sim 10^8 - 10^{13} \times$ OZMA			

for artifacts at the stable libration points of the Earth–Sun and Earth–Moon systems, with both optical and radar systems. People have looked for infrared lasers, and for Dyson spheres in the infrared (civilizations that surround themselves with a cocoon so they can use all their sunlight), for molecular lines in the millimetric regime, and for optical lines characteristic of fission products (from civilizations that dump their radioactive and other wastes into their sun, partly to get rid of them, partly to signal to us).

I wish now to summarize the handful of contemporary searches and the two or three searches that are planned.

There are four significant searches in progress. At Arecibo, the Berkeley group (Stu Bowyer, Dan Werthimer, and crew) is running an ingenious search called SERENDIP, a strained acronym standing for something or other. It uses the Arecibo dish in an unusual parasitic mode – namely, by using the wrong feed carriage. So the beam does not track the celestial sphere, but instead describes funny little curlicues in the sky (while the primary user on the other feed carriage does track a single source position). In a year's time they cover almost the entire

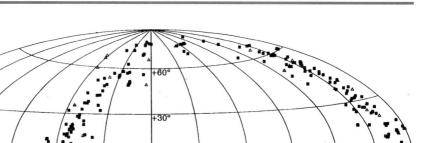

Galactic Map of Candidate Events as of September 1994

■ – Events seen on 2 different days (barycentric reference frame)
▲ – Events seen on 2 different days with 2 detections per observation
    (no reference frame assumptions)

sky visible from Arecibo; in fact, many positions are observed three, four, or five times. The hardware is a four-megachannel Fourier spectrometer, observing at 430 megahertz. At this frequency one gets lots of interference.

FIGURE 9.11
Candidate events from two years of SERENDIP III's parasitic operation at Arecibo.

The idea is to take these seventy trillion signals coming in, and you look for regularities in frequency and space and time, rejecting everything that smells like a rat: signals that stay on too long while the antenna's moving, signals that keep coming back at the same frequency because they're from some local transmitter, and so on, and in the end you wind up with a residuum of a few hundred interesting events. Figure 9.11 shows the result of two years of running, about two hundred fifty events that were seen on more than one occasion. The statistics happen to be just at the level of chance, and none of these has been seen three times (a single such instance of which would be extremely improbable). This project was done at low cost and with great elegance of design and execution.

Meanwhile, at Ohio State University (the antenna that looks like a football field), the longest-running search has just increased its channel count by a factor of 100,000, with the receipt of a four-megachannel system from Berkeley. A photograph of their upgraded receiver is in Figure 9.12, sitting in the leftmost rack of the obligatory four racks of equipment. Perhaps the most tantalizing candidate signal in SETI came from Ohio's search, the celebrated WOW signal, which displayed an uncanny match to the antenna's beam profile as it drifted through.

FIGURE 9.12
Four-megachannel
Berkeley
spectrometer at
Ohio State
University
observatory.

This event has now occupied hundreds of observing hours in futile attempts at reacquisition. These occasional wonderful signals have been seen by just about every search, invariably operating in a mode that makes immediate reobservation impractical. At the end, I will comment on what should be done about things like that.

Here's another ongoing experiment (Figure 9.13), the infrared apparatus of Charlie Townes and Al Betz, looking for carbon dioxide laser beacons. Charlie has always favored infrared and thinks the rest of us are missing out on a good bet. This is elegant physics – they use true coherent detection, in an optical heterodyne system.

Finally, our search (META) at Harvard and Buenos Aires, sponsored primarily by The Planetary Society. My son Jake (Figure 9.14), when he was six years old, is pointing to the Cassegrain radome of the equatorial 84-foot dish at Harvard, Massachusetts. Our apparatus is shown in Figure 9.15, with its eight million channel analyzer, turned on in 1985 with some fanfare (Figure 9.16). What you are seeing, from left to right, is 1) the switch that would turn it on if anybody had a hand on it; 2) the fellow who gave the money to build the stuff that the switch turns on; 3) the fellow who talked this guy into giving the money, and 4) one of the guys who built it.

A second identical system (META II) was built by the Argentines, who shipped it to their 30-meter dish, where it is performing a search of the southern sky, and also a set of simultaneous observations of the portion of the sky visible from both observatories.

What have we found with these searches? Things like the WOW event, things that go bump in the night, things that never come back.

FIGURE 9.13
Ten-micron infrared heterodyne SETI at Mount Wilson.

For instance, in Figure 9.17 is the result of the META search at Harvard, 10^{13} channels examined, during five years of continuous observations. Strong events that survive all tests for fishiness are plotted in this sky plot (reported in *Astrophysical Journal*, 415, 218 [1993]); the larger points are far too strong to be due to chance, whereas the smaller ones are consistent with the statistical noise tail. What may be rather interesting is the clustering of the five strongest signals in an apparently nonrandom arrangement relative to the galactic plane. We've spent many, many days on each one of these, in an attempt at reobservation, and we haven't gotten any to come back. It's like the cargo cults – we've deployed telescopes, and we'd do anything (even build airstrips!) to make those signals come back.

What can one conclude from three decades of negative results in SETI? Of course, it would be much better to have found something – we in the SETI business all wish we had magical results like those that Ed Stone showed this morning from the spectacularly successful program of planetary exploration. But in SETI you either get totally magical results or nothing at all. Carl and I made an effort to extract the most from the "nothing at all" of our search, and it goes something like this: If you believe that there may be supercivilizations, in the Kardashev sense (ones that extract all the energy of sunlight falling on

FIGURE 9.14
Fully steerable
84-foot dish, doing
full-time SETI at
Harvard.

their planet – Type I; or all the energy radiated by their sun – Type II),
and if you confine your attention to those that continuously transmit
radio carriers in the bandpass of our search, using a significant fraction of their power resources, then the negative results of META rule
out Type I civilizations within 700 parsecs, if they transmit isotrop-
ically, or anywhere in the galaxy, if they transmit in our direction
with 30 decibels of antenna gain. Type II civilizations are ruled out,

FIGURE 9.15
Eight-megachannel META SETI at Harvard.

under the same ground rules, clear out to 22 megaparsecs, even with isotropic antennas (this includes the local group and the Virgo cluster). Even civilizations as primitive as ours are ruled out, to 7 parsecs (isotropic transmission), or anywhere in the galaxy (70 decibels of antenna gain). Note, however, how restrictive our initial assumptions are – continuous carrier transmission, at our choice of frequency, beamed at us if directional antennas are used.

Another way to put it is to say that SETI, to date, has barely scratched the surface. We'd like to search in a way that has a significant chance of finding beacon transmissions not so narrowly constrained. And that brings us, finally, to the searches now in planning and construction.

The SERENDIP folks, who parasitically use the Arecibo antenna, are gearing up to 120 megachannels (SERENDIP IV), to cover Arecibo's sky with greatly increased bandwidth.

At Harvard we're building a two-beam transit system (Figure 9.18), with a pair of east–west feedhorns at the focus of our dish; a third low-gain antenna serves as a terrestrial veto. The hope is to force any genuine transitting source in the sky to run the gauntlet, and therefore to give us robustness against things that go bump once in the night. With this system (BETA) a signal has to go bump twice, in the right way (east, then west, and never terrestrial). The basic processor is a

FIGURE 9.16
Celebrities
blocking view of
the 84-foot dish.
(Courtesy of
K. Beatty, Sky
Publishing,
Cambridge, MA)

four-megachannel board, does a 4 million point complex Fourier Transform in 2 seconds, about 300 million instructions per second. It's a cousin of SERENDIP. A bunch of them stuck into a rack are in Figure 9.19, along with the down converters and digitizers; you're looking at a 250 million channel receiver, about 20 billion operations per second. This system is going on the air in 1995.

In connection with these 100+ megachannel systems, I'd like to quote Frank Drake. In 1985, on the twenty-fifth anniversary of Ozma, Frank said "Today we're doing the eight million channel systems with very high sensitivity, and sure enough, twenty years from now, we will have a hundred million channel system or so." Well, it's been just nine years and we've already got a quarter of a billion. As Bohr once said, "It's difficult to predict, especially the future."

Saving greatest for last, the NASA project has now been reborn (Phoenix), with senator-proof private funding. This project is *really* big, with boxes that barely fit in very large cargo airplanes. The system knows how to recognize fragmentary pulse trains, chirps, and various combinations, in addition to bland carriers; and it will search a wide band – 2 gigahertz or more – with the sensitivity of the largest antennas on earth.

To indicate how things get better with time, a comparison that Kent Cullers worked out, using a straightforward metric of search power, is

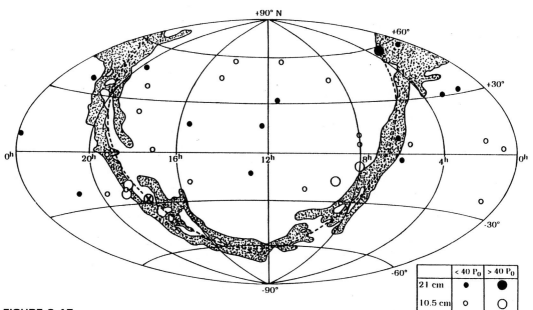

FIGURE 9.17
Surviving events from 5 years of META. Filled circles are 1.4 GHz, open circles are 2.8
GHz; the five strongest events are shown as larger circles. The center of the Milky Way is
indicated by an *X*.

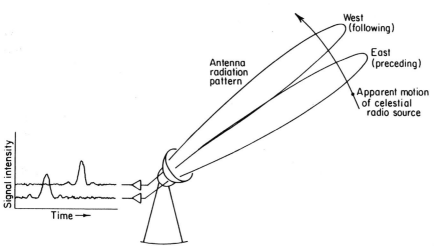

FIGURE 9.18
Dual-beam transit
SETI. A third
channel, fed from
a low-gain all-sky
discone, provides
a veto.

shown in Table 9.2. With a table like this we can outrun even Frank's
optimistic projections. Kent even used Drakian units – ozmas and oz-
mas/second (Ozma kept Frank busy for 2 months, back in 1960) –
listed as logarithms, because the gains have been so spectacular. You
can see that the current and future systems, depending on the kind of
signal that you posit, are something between seven and fourteen orders of magnitude improved over Ozma in sensitivity. That's not bad
for thirty years.

FIGURE 9.19
Two hundred fifty-million-channel spectrometer for SETI. BETA operates at 40 billion operations per second, contains 3.4 gigabytes of RAM serving over 200 processors, and produces 250 MByte per second of spectral data.

A last comment about these improving searches: we've all learned from these once-a-month or once-in-the-middle-of-the-night events that we've got to include much better schemes for the mitigation of radio frequency interference, occasional errors in the processors, and so on; and we have to include means for quick follow-up, whether there's an operator at the controls or not. All of us pursuing the SETI enterprise now know this well. Just to give an example, in our new BETA system we've implemented a system with three horns, so that the signal has to display proper sidereal drift through the sky horns while remaining undetected in the terrestrial antenna; BETA also uses redundant processors, and it uses extensive parity checking. If a signal passes these tests, the system responds by leapfrogging the antenna

TABLE 9.2
COMPARATIVE SEARCH SENSITIVITY AND RATES FOR
CONTEMPORARY SETI

Comparative Sensitivity (for one direction in the sky)			
Search Size (log ozmas)			
Search Name	Nondrifting CW	Drifting CW	Nondrifting Pulses	Drifting Pulses
---	---	---	---	---
META	8.96	6.36	8.31	6.36
SERENDIP	7.56	7.56	7.56	7.56
OSU	5.51	5.51	5.51	5.51
BETA	9.41	9.41	9.41	9.41
OPTICAL AST	4.27	4.27	4.27	4.27
RADIO AST	6.31	6.31	6.31	6.31
NASA	13.60	13.60	14.92	14.92

Comparative Search Rates			
Search Rate (log ozmas/sec)			
Search Name	Nondrifting CW	Drifting CW	Nondrifting Pulses	Drifting Pulses
---	---	---	---	---
META	7.66	5.06	7.01	5.06
SERENDIP	7.56	7.56	7.56	7.56
OSU	5.51	5.51	5.51	5.51
BETA	9.41	9.41	9.41	9.41
OPTICAL AST	4.27	4.27	4.27	4.27
RADIO AST	6.31	6.31	6.31	6.31
NASA	11.22	11.22	12.23	12.23

westward in hour angle, inviting the source to perform an encore transit. Future SETI will see increasing use of such gimmicks.

Finally, a personal note to Carl: You set the stage with your landmark book (with Shklovskii), *Intelligent Life in the Universe*, for a generation of searchers. It was that (which you used as course notes at Harvard in the 60s, and which I took vicariously through my roommate without paying any tuition), together with the influence of giants like Ed Purcell and Frank Drake, that drew me into SETI, where I seem to be stuck ever since. SETI is alive today, thanks both to your behind-the-scenes advocacy and to public advocacy such as the petition that you organized for *Science* magazine. Your effectiveness in science education, in the form of The Planetary Society leadership, in the forum of public education (which helps generate an enthusiastic public that's willing to shell out the bucks for this), and also in your patient education of senators, has been absolutely crucial to SETI. You have, and very much so, had your "hand on the switch" (Figure 9.20). Thank you, and happy birthday!

FIGURE 9.20
Courtesy of
K. Beatty (Sky
Publishing,
Cambridge, MA)
and John W.
Forbes, Jr.

Note added in proof: In the two years since this symposium the BETA and Phoenix searches have become fully operational, and SERENDIP IV awaits the upgraded Arecibo; that was expected. The unexpected is the discovery of nearly a dozen extrasolar planets, some circling stars like our own; a meteorite from Mars that may contain fossil life; and new evidence for liquid water within Jupiter's moon Europa. SETI is more exciting than ever before, now sadly without Carl's dynamic participation.

— 10

Do the Laws of Physics Permit Wormholes for Interstellar Travel and Machines for Time Travel?*

KIP S. THORNE

California Institute of Technology

Our friend and honoree, Carl Sagan, is not only a fine fellow, a great scientist, and an outstanding communicator; he is also a talented novelist, as those of you who have read his book *Contact* [Sagan 1985] must know.

Now, it is a rare and perhaps unique happening that a science fiction novel like *Contact* generates an important new direction in scientific research. But then, Carl is a unique individual. I will describe today how, through *Contact*, he has triggered a community of theoretical physicists to study some extreme warpages of spacetime that they previously had steered clear of, and how those studies are producing new insights into the nature of space and time.

It all began about nine years ago with a telephone call from Carl to me, in which Carl said, "I've just finished writing a novel about the first human contact with an extraterrestrial civilization, and I want to be absolutely sure that I've got all of the general relativity right." That is typical of Carl; he wants to get it right. Completely right.

So Carl sent me the manuscript of his novel – a fascinating story in which the heroine, Eleanor Arroway, travels through a black hole (an extreme type of space warp) to the center of the galaxy, spends a day there, and then, passing through some sort of time warp, returns to Earth at the very moment she departed. So there was the challenge: Could this be made scientifically respectable? Respectability came easily, as it turned out. Carl had foreseen remarkably much, despite not being an expert in Einstein's general relativity – Einstein's theory of gravity and spacetime warpage.

*For a more detailed account of this topic at roughly the same level as this lecture, see the last chapter of *Black Holes and Time Warps: Einstein's Outrageous Legacy* [Thorne 1995]. For a more technical overview, see [Thorne 1993].

Now, the technology to manipulate spacetime warps in the manner Carl had envisioned is as far beyond our capabilities as space travel was beyond the Neanderthals. Of course Carl recognized this, so in his novel he had an extremely advanced civilization place the technology in our hands, via a long radio message that Eleanor Arroway receives and decodes.

Now, I can't predict what an extremely advanced civilization will be capable of doing or capable of teaching us to do. But I can ask myself, "What do the fundamental laws of physics allow?" Suppose that this civilization is infinitely advanced. Suppose it is constrained only by what the fundamental laws of physics prevent it from doing, and nothing else. Then, is it possible to create and manipulate such spacetime warps?

As I began to ask this question, it quickly became evident that here was a powerful way to probe the fundamental laws of physics. Ask not what occurs in nature, nor what we humans are capable of doing. Ask, instead, *What constraints do the laws of physics place on an infinitely advanced civilization?*

There is a precedent for questions like this: In the early twentieth century, when Albert Einstein was developing his general theory of relativity, humans had little hope of measuring the warpage of spacetime on which his theory was based. Real, quantitative tests would become possible only in the 1970s, more than fifty years after he formulated his theory, so he could not rely on real experiments as a guide. Instead, he used thought experiments – experiments in which he imagined doing things like jumping on a beam of light and riding with it at nearly the speed of light, which we really can't do because of our puny technology. By thinking about such hypothetical experiments and computing what the laws of physics say should be the outcome, Einstein gained insight into the fundamental laws, insight that guided his creation of relativity theory.

Similarly today, in trying to go beyond general relativity – in trying to understand how gravity behaves in the cores of black holes or in the big bang singularity where the Universe began – we physicists enter a domain where we cannot do real experiments, and so we use thought experiments. But we have shied away from what is perhaps the most powerful kind of thought experiment of all, what in my circles has come to be known as a Sagan-type question. These are thought experiments that ask, "What constraints do the laws of physics place on the activities of an infinitely advanced civilization?" We haven't asked such questions, until now, because they seemed too much like science fiction and thus seemed somewhat unrespectable in the staid science community.

We now realize, however, that if we ask what an infinitely advanced civilization can do, we are asking ourselves about the fundamental laws of physics in a very deep sort of way. And thus it is that,

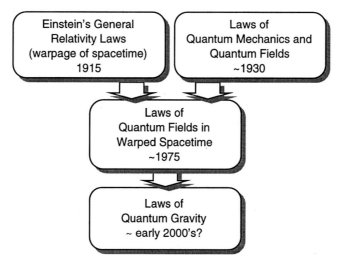

FIGURE 10.1
The laws of physics that are probed by Sagan-type questions.

triggered by Carl's challenge to me, we have begun to ask Sagan-type questions.

Let me say a few words about the physical laws that we have been trying to probe in this way.

In the twentieth century, there were two revolutions that brought us two new sets of physical laws (Figure 10.1): *Einstein's 1915 general theory of relativity*, in which he told us that spacetime can be warped by dense concentrations of matter and energy and that this warpage manifests itself in part as gravity; and *the laws of quantum mechanics and quantum fields*, dating from the 1920s and 1930s, which are the laws that govern atoms, molecules, particles of light (photons), and other entities on very small scales.

In the half century since these revolutions, it has become clear to us that, underlying general relativity and quantum mechanics there must be a set of unified laws. In these unified laws, spacetime warpage, which governs things in the large, must merge with quantum mechanics, which governs things in the very small. Spacetime warpage and quantum mechanics must come together, forming a new set of laws called *quantum gravity*, and these new quantum gravity laws must govern what goes on at the center of a black hole and in the big bang singularity where our Universe was born.

Theoretical physicists have been struggling since the 1950s to discover the true nature of these quantum gravity laws, but the struggle has been extremely frustrating, with only occasional whiffs of success. There was, however, a partial triumph around 1975, when a handful of theorists – among them Leonard Parker, Bryce DeWitt, Stephen Hawking, and Robert Wald – formulated a partial marriage, one in which the electromagnetic field and the neutrino field and other fields are fully governed by quantum mechanics and live in Einstein's warped spacetime. Unfortunately, the resulting *laws of*

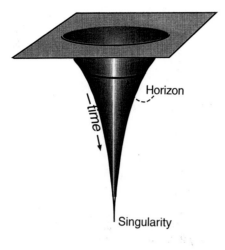

FIGURE 10.2

Depictions of the warpage of space around and inside a black hole, with one spatial dimension suppressed.

quantum fields in warped spacetime continue to treat spacetime itself as classical, that is nonquantum-mechanical. The fields themselves behave in probabilistic ways characteristic of quantum mechanics, but the spacetime is not probabilistic at all; it has a very definite form and shape. And thus, the full laws of quantum gravity still elude us.

As I shall describe, in the last several years we have been using Sagan-type questions to probe the full range of spacetime-warpage laws (Figure 10.1): general relativity (which we understand very well), the laws of quantum fields in warped spacetime (which we understand moderately well but not fully), and the laws of quantum gravity (which still largely elude us).

In Carl's novel, he had his heroine travel through a *black hole*. A black hole is an extreme type of spacetime warp. It's an object that is not made from matter, but instead is made solely from a warpage of space and a warpage of time. It is a three-dimensional analog of what I show in Figure 10.2 in two dimensions.

Imagine that we live in a two-dimensional universe, so we are like ants who can crawl around on the surface shown in Figure 10.2. That surface is our whole universe. Now, we're blind ants, so we can't look across from one side of the surface to the other and see that it is embedded in a surrounding three-dimensional space. However, crawling around on the surface we can measure the circumference of a circle and then its diameter and can discover that the diameter is much larger that the circumference divided by π, and thereby we can discover that the space of our universe is warped.

Now, a black hole has a surface, which is called its *horizon*. In Figure 10.2, the horizon is drawn as a circle, since the dimensionality of space has been reduced by one: Space itself looks two-dimensional rather than three-dimensional; the horizon is a one-dimensional circle rather than a two-dimensional spherical surface.

The laws of general relativity insist that the horizon is a one-way membrane. Things that fall inward through the horizon can never ever reemerge. In essence, this is because of the hole's warpage of time. Inside the horizon, time is flowing inexorably downward, toward a singularity that lurks at the hole's center; and we humans are forced always to travel forward in time, which means downward from the horizon, inexorably into the singularity. To do otherwise – to thwart the direction of time flow – would require an infinite outward acceleration, which the laws of physics forbid.

This time flow seals the fate of Eleanor Arroway, Carl's heroine. If she plunges into a black hole, she will be pulled inexorably downward, into the singularity.

The nature of a black hole's core was not clearly understood by most scientists and nonscientists, when Carl wrote his novel. There had been lots of articles in the popular literature and some in the technical literature claiming that you could travel through the hole's core and reemerge elsewhere in the Universe. Well, you can't. The downward flow of time prevents you from ever turning around and reemerging through the horizon; and it drags you into the singularity, which is an impenetrable barrier against onward travel.

Now, the singularity is a place where spacetime is warped so extremely strongly that it is governed not by general relativity but instead by the laws of quantum gravity. Since we don't know those laws, you might hope that they would permit Eleanor Arroway to survive and reemerge elsewhere in the Universe. Not so. As Eleanor falls in, before general relativity fails and quantum gravity takes over, the warpage of spacetime has crushed her to a density far greater than the billions of tons per cubic inch that characterizes matter inside neutron stars, and she is very, very dead. This is not the fate that Carl envisioned for Eleanor, particularly since I think his daughter was the model on which Eleanor was molded.

Evidently, Carl's novel needed a bit of changing. The black hole needed to be replaced by some other type of spacetime warp, through which Eleanor could travel and survive. A *wormhole* was the obvious choice – obvious to me, though not to people outside the relativity community since at that time wormholes were obscure, hypothetical objects that only relativity afficionados were aware of.

A wormhole has two *mouths*, each one a spherical surface somewhat like the surface of a large rubber ball, except it is not made from rubber but instead from a warpage of space. If you stick your hand into one of the mouths (one of the spheres), you might see your fingers emerge from the other mouth, even though it is 20 feet away on the opposite side of the room. At least, this is what might happen if the wormhole mouths were the size of basketballs and the wormhole itself (the connection between the mouths) were short, say an inch or two in length.

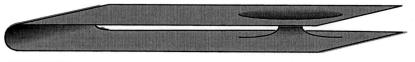

FIGURE 10.3

A wormhole that connects two widely separated regions of space in our Universe.

In essence, the wormhole is a handle in the topology of space. It provides an unexpected way for your fingers to travel across the room: through the wormhole (the handle; a distance of an inch or two) rather than through normal space (a distance of 20 feet).

You can understand this from the two-dimensional analogy of Figure 10.3. Our Universe is like the large two-dimensional surface shown there. We, being blind ants, cannot see that our Universe is bent around in the surrounding hyperspace. (In the bending region – the left edge of Figure 10.3 – we will measure circumferences of circles to be π times diameters, just like normal, and thus will infer that our space there is flat.) However, we can discover that there is a short wormhole connecting two different locations in the Universe: we can travel through the wormhole, using it as a shortcut to get from the top mouth to the bottom mouth.

Wormholes were discovered, amazingly, as a solution of Einstein's general relativity equations in 1916 by Karl Swarzschild [1916], but it required a reinterpretation of Swarzschild's mathematics by Ludwig Flamm [1916] in Vienna in that same year, to reveal the wormhole nature of Schwarzschild's solution. This wormhole solution was long doomed to obscurity because it was so bizarre that people were loath to consider it as something that might exist in the real Universe. Nevertheless, John Wheeler and a number of other theorists over the years did investigate the properties of such wormholes mathematically.

Among these theorists, it was Martin Kruskal who, in 1959, discovered a fundamental problem with the Swarzschild–Flamm wormholes: If you want to travel through one, you can't, because as time passes the wormhole's throat pinches off. It pinches off so quickly that, while you're trying to travel through, you get caught, crushed, and killed.

So, when Carl asked me to help make his novel scientifically respectable, the question arose of how do you hold a wormhole open so that Eleanor Arroway can travel through it from Earth to an orbit around the star Vega. And that was where issues in fundamental theoretical physics arose.

Playing around with Einstein's equations, Mike Morris (a student of mine) and I [Morris and Thorne 1988] quickly deduced that, to hold the wormhole open, you must thread it with a new type of matter, something that we have chosen to call *exotic matter*, but that in the standard jargon of theoretical physics is "a material that violates the

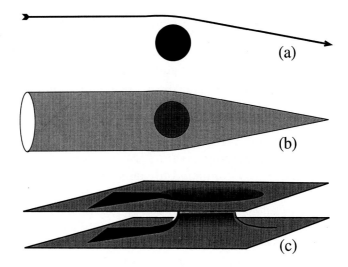

FIGURE 10.4

(a) A light ray from a distant star gets deflected by the Sun's spacetime warpage. (b) The Sun's warpage focuses a giant laser beam. (c) A wormhole's spacetime warpage defocuses a laser beam.

averaged null energy condition." I use this horrible phrase to show you that I can be respectable; I can mutter the jargon along with the best of my colleagues (and I do so in technical papers). But I prefer to call it exotic matter.

The following thought experiment shows why exotic matter is needed to hold a wormhole open:

When a light ray from a distant star passes near the Sun and on to Earth, the sun's gravity (its spacetime warpage) deflects the ray as shown in Figure 10.4a. Similarly, it seems reasonable that, if somebody with a gigantic laser beam were to shine it through and around the Sun, the beam would get deflected on all sides and would thereby get focused as shown in Figure 10.4b. What is it that causes that focusing? It's the warpage of spacetime around the Sun. That warpage in turn is produced by the energy or mass inside the Sun (mass and energy are equivalent according to Einstein) and also by the pressure along the direction in which the beam travels. It turns out to be the sum of the energy density and the pressure in the Sun that produces the warpage which focuses the laser beam.

A wormhole, like the Sun, can focus light rays. Suppose we send a laser beam radially into one mouth of the wormhole, as shown in Figure 10.4c, where one dimension is suppressed. The circular symmetry of the diagram (the spherical symmetry of the real wormhole, with the suppressed dimension restored) guarantees that the laser beam will keep traveling radially as it enters, traverses, and emerges from the wormhole. As a result, the light rays that enter the wormhole converging, emerge diverging. The laser beam, instead of being focused as by the Sun, gets defocused. The wormhole does the opposite to what the Sun does, and so its warpage must have the opposite sign, and the thing that causes the warpage must have the opposite sign. The wormhole's energy density plus pressure must be negative rather than positive. In the jargon of physicists, the wormhole must be threaded by "material that violates the averaged null energy condition"; in my language, it must be threaded by exotic matter.

Is exotic matter possible, according to the laws of physics? Many physicists have taken it for granted that the answer is no; exotic matter should be impossible. Why? Because under ordinary circumstances pressure is tiny compared to energy density,[1] and if the pressure is tiny, then exotic material must have a negative energy density; and negative energy seems inherently nasty to physicists. But we must not let our prejudices cloud our reason. Black holes also once seemed nasty, yet we now are convinced they do exist in our Universe. The rational thing to do, then, is to ask the fundamental laws of physics whether or not they permit exotic matter.

This question has become an important topic of theoretical physics research as a result of Carl's novel. And its importance has been enhanced by theorists' recognition that exotic matter is not only the key to holding wormholes open, it is also a key to making time machines and to the nature and perhaps even existence of singularities [Thorne 1993].

So, do the laws of physics allow exotic matter, and allow it in enough quantity to hold a wormhole open? The final answer is not yet in. Gunnar Klinkhammer, a graduate student in my Caltech group, was the first to tackle this question in response to Carl's novel. Klinkhammer [1991] managed to prove rather quickly, using the laws of quantum fields, that when spacetime is very nearly flat, hardly warped at all as in Earth-bound laboratories, then you cannot have exotic matter at all.

Then, switching to the new laws of quantum fields in warped spacetime, Klinkhammer discovered a particular example of a special spacetime warpage in which exotic matter can exist [Klinkhammer 1991]. Thus, in flat spacetime exotic matter is forbidden, but in warped spacetime – or, at least in one warped spacetime – it is allowed.

Klinkhammer's warped example is fairly easy to explain. Suppose that space were so badly warped that, when you walk into the east wall of your bedroom, you immediately find yourself coming out of the west wall. In other words, your bedroom is closed up in the east–west direction. It's as though you lived on the surface of a cylinder, and in walking into the east wall and emerging through the west and returning to where you started, you have traveled around the

---

[1]More specifically, pressure is small compared to energy density when, as we must in this context, we include the mass of the material as part of its energy in accord with Einstein's equation $E = Mc^2$.

cylinder. Klinkhammer showed that, if you put an ordinary electromagnetic field into that kind of warped space, and you then remove
all the energy you can from the electromagnetic field so it is in what
we call its vacuum state, then the little bit of irremovable, quantum-
mechanical, probabilistic, fluctuational behavior that remains in the
electromagnetic field – that little bit is exotic. These *electromagnetic
vacuum fluctuations*, as they are called, would have vanishing energy and pressure in the flat spacetime of everyday experience, but
Klinkhammer's warped spacetime endows them with negative energy
density and negative pressure, so they violate the averaged null energy
condition. They become exotic.

This tantalizing mathematical example of exotic matter has been
reenforced by Robert Wald and Ulvi Yurtsever [1991] at the University
of Chicago. They have shown that a very wide, and indeed generic,
set of spacetime warpages endows vacuum fluctuations with exotic
behavior. Not every spacetime warpage will do so, but a huge number will.

These beautiful results are only the first steps in what is turning out
to be a painful, long struggle to try to understand whether wormholes
can be held open. Sometimes answers come quickly. Sometimes they
come very slowly and with great effort. This is the slow kind.

What we want to know is whether vacuum fluctuations, which are
made to behave exotically by the spacetime warpage in which they
live, can generate – via their energy and pressure – the very warpage
that is allowing them to be exotic. And we want to know whether this
can be done with sufficient vigor to hold a wormhole open. We've
got to have it going both ways. The warpage must make the vacuum
behave exotically, and the vacuum must feed back and produce that
very warpage that is enabling it to be exotic.

We don't yet know the answer, so I'll just tell you my guess. My
guess is that the loop can be closed, and you can have wormholes that
can be traveled through – if you are infinitely advanced. But this is
just a guess. The true answer is coming so slowly that it is unlikely to
be in hand by Summer 1997, when the film version of *Contact* hits the
theaters. Thus, Carl is probably safe invoking wormholes and exotic
matter in the film, as he did in the final version of his novel.

About the time that Mike Morris and I wrote our first technical
manuscript on wormholes and the challenge of holding them open
[Morris and Thorne 1988], Tom Roman at Central Connecticut State
University startled us with this insight: If an infinitely advanced civilization can make wormholes and hold them open, then it should
be only a minor further step to turn such wormholes into time machines. Minor for an infinitely advanced civilization that is, not for
twenty-first-century humans.

One way to do so involves a thought experiment in which my wife
Carolee and I are both infinitely advanced. We hold hands through a
small-mouthed wormhole, she outside one mouth and I outside the

other. I remain at home with my mouth in the comfort of our living room while Carolee journeys off with her mouth at nearly the speed of light, into the Universe and back. Despite her distant travel, the length of the wormhole, from her mouth to mine, remains only a few inches and we continue holding hands through those inches.

As Carolee travels, we both look at her watch through our wormhole mouths. The watch appears to both of us to be ticking at a normal rate. There is no time warpage inside the wormhole, relative to either of us.

Carolee's journey takes 1 hour, from 3 P.M. to 4 P.M. on a Saturday afternoon, as experienced by her and as ticked by her watch. When she returns after 1 hour (an hour on which we both agree, looking through the wormhole at each other), she tells me through the wormhole that she is back, so I leave my living room and wormhole mouth, and go out to the family spacecraft port to greet her. She's not there.

Why not? Because as seen in the external Universe (i.e., not through the wormhole), time has flowed at a different rate on Earth that in Carolee's near-light-speed spacecraft. In the spacecraft port I must await her return until 4 P.M. the next day, Sunday. This discrepancy of time flow as seen in the external Universe is an unavoidable consequence of Einstein's relativity theory. It goes under the name twin paradox because in the standard story without a wormhole the two protagonists are twins and she ages just 1 hour during the trip while I (temporarily her twin) age 25 hours awaiting her return.

At 4 P.M. on Sunday when Carolee finally returns to the spacecraft port, I greet her and see her holding hands with somebody through her wormhole mouth. That somebody is me, at 4 P.M. the previous day, Saturday. On Carolee's side of the wormhole it is Sunday (though she only aged 1 hour to get there). On the other side, as seen through the wormhole it is Saturday, with just 1 hour having passed.

The wormhole has now become a time machine. By climbing through it from her mouth to mine, she can travel from Sunday 4 P.M. to Saturday 4 P.M. If I had climbed through it from my mouth to hers before going out to the spacecraft port, I would have entered at 4 P.M. on Saturday and emerged at 4 P.M. on Sunday. Her journey has converted the wormhole into a time machine, with a 1-day time difference between the two mouths.

How do I know this? It is an inevitable consequence of the fully understood general relativity laws. It does not depend on the moderately well-understood laws of quantum fields in curved spacetime (except that those laws are needed to tell us how to make the exotic matter that holds the wormhole open). And it does not depend at all on the ill-understood laws of quantum gravity.

Or so I thought for several years. But I get ahead of myself.

When my students and colleagues and I pursued these lines of thought a bit further, it turned out that, according to general relativity, it is embarrassingly easy to make a time machine, or in technical jargon to make *closed timelike curves*. Embarrassingly easy if you're an infinitely advanced civilization, but utterly impossible if all you have is twenty-first-century technology.

In the technical literature, there now are many examples of closed timelike curves (time machines), besides moving wormholes. In one example, devised by the great mathematician Kurt Gödel [1949], the Universe is set spinning at a high rate and is endowed with a cosmological constant, and it thereby becomes a gigantic time machine. By traveling out through such a Universe at high speed on a carefully chosen path, you could return before you started, and you would do so without the aid of any wormhole. Recently, following up on wormhole examples, Richard Gott [1991] of Princeton University showed how to use cosmic strings (cracks in the fabric of space) to make time machines. And recently, Stephen Hawking of Cambridge University showed how a spinning and slowly contracting star that contains some exotic material can become a time machine [Hawking 1992].

These examples are embarrassing because as physicists we would prefer to believe that backward time travel is impossible. Why? Because it seems one could go back and change history, and that would create problems for theoretical physics. As physicists, we could no longer pose initial conditions and straightforwardly evolve them into the future using standard laws of physics. Our evolutions would seem to produce paradoxes. (Whether this is really so has become a major topic of research, one that I don't have time to describe today [Thorne 1995, Thorne 1993].)

In fact, as Hawking likes to argue, only half with tongue in cheek, there is experimental evidence against backward time travel: We have not been invaded by hordes of tourists from the future; at least we think we haven't.

With tongue removed from cheek, Hawking [1992] has recently proposed a *Chronology Protection Conjecture* (then restoring his tongue to his cheek, he has characterized his Conjecture as "making the world safe for historians"). Hawking's Conjecture states that the laws of physics must somehow conspire to make backward time travel impossible and must do so not just for wormhole-based time travel but also for any other time machine that an infinitely advanced civilization ever tries to construct.

Hawking's conjecture, and a vigorous effort to test it, are part of the legacy of Carl's novel. Without that novel, our community would probably not have ventured down these intellectual paths.

Sung-Won Kim and I [Kim and Thorne 1991] and Valery Frolov [1991] have identified a physical mechanism that we and Hawking suspect will always enforce Chronology Protection. This mechanism, first discovered in a different context by Bill Hiscock and Deborah Konkowski of the University of Maryland [Hiscock and Konkowski

1982], appears to be universal. Whenever one tries to make a machine for backward time travel, this mechanism steps in and destroys it just before it begins to operate. And it does so whether the time machine is made from wormholes or cosmic strings or spinning stars or the entire Universe. They all must explode self-destructively, it seems, when one tries to turn them into time machines.

To understand this mechanism, imagine Carolee bringing her wormhole mouth back toward Earth, near the end of her high-speed journey. During her return trip, there comes a first moment when she can travel back in time and meet her younger self. To do so at that first possible moment, Carolee in her spacecraft must jump into her wormhole mouth, emerge from my mouth at home on Earth, and then travel out from Earth as fast as possible, so as to reach the spacecraft just before she jumps through the wormhole.

Now, as fast as possible means at the speed of light, and Carolee can't travel quite that fast, even if she is infinitely advanced. The laws of physics forbid it. However, light can travel that fast, as can electro-magnetic vacuum fluctuations. Thus, light and vacuum fluctuations are the first things that can travel backward in time through the worm-hole and meet their younger selves.

The vacuum fluctuations, we believe, are the culprit that destroys the wormhole at the moment it first becomes a time machine. They do so by traveling around and around – through the wormhole, then out through the Universe from the Earth to the spacecraft, then through the wormhole again and out through the Universe to the spacecraft again – piling up on themselves not just in space but also in time. They do this right at the moment when they are first able to time travel. Returning from its first round trip, each fluctuation piles up on itself in spacetime, thereby doubling its strength. Returning from its second trip, it piles up on itself again, tripling its strength, and so on. The fluctuational power grows infinitely and destroys the wormhole precisely at the moment when time travel is first possible.

In 1990, when Kim and I saw this conclusion emerge from our cal-culations, we were amazed. We had expected the circulating vacuum fluctuations to behave like circulating light (Figure 10.5a). The worm-hole acts like a diverging lens and so, when a beam of nearly parallel light rays goes into Carolee's wormhole mouth, it must emerge from mine splaying out in many directions, so only a bit of the light can return to her mouth and pile up on itself – so small a bit that no explo-sion is created at all. It was obvious that this was how ordinary light would behave, and we expected the same of vacuum fluctuations.

But Bill Hiscock warned me not to be so confident. And he was right. The vacuum fluctuations do, indeed, splay out of my worm-hole mouth just like light; but then, according to our calculations, the fluctuations focus themselves back down onto Carolee's mouth, with no losses at all, and pile up on themselves (Figure 10.5b). This was the unavoidable prediction of the laws that govern the vacuum

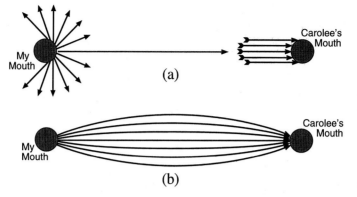

FIGURE 10.5
(a) The behavior of light rays that travel through the wormhole at the moment it becomes a time machine. (b) The behavior of vacuum fluctuations.

fluctuations, the laws of quantum fields in curved spacetime; but I still can't make the prediction seem physically plausible because we still don't understand those laws fully.

We don't understand in physical terms, but we can calculate. And our calculations show that this refocusing and infinite pileup of vacuum fluctuations is universal. The fluctuations must focus, pile up, and create an explosion whenever an infinitely advanced civilization tries to create a time machine, and no matter how the civilization tries to do so: with wormholes or cosmic strings or stars or the whole Universe or whatever. Moreover, the vacuum fluctuations cannot be shielded out. They will penetrate anything one places in their way. So their pileup and explosion does, indeed, seem an attractive way to keep the world safe for historians.

Attractive, but not certain. We are not absolutely sure, as yet, whether the explosion triggered by these fluctuations is *always* strong enough to destroy an incipient time machine. My colleagues and I have gone back and forth over this for several years. Sung-Won Kim and I initially thought the time machine might survive, at least if it is based on wormholes. I'll explain why in a moment. A year later, Hawking said to us, "You're all wet, you made a mistake, the explosion will always destroy the time machine." Hawking's argument convinced us, so we changed our manuscript just before it was published [Kim and Thorne 1991]. Then Hawking's student James Grant wrote a paper with Hawking's blessing, saying that Hawking was wrong, the explosion sometimes will not destroy the time machine. Then I found a flaw in Grant's argument and convinced him and Hawking that the time machine will be destroyed – I convinced them just in time for Grant to change his manuscript before publication [Grant 1973, Thorne 1993].

Why all this equivocation? Because in our calculations, just as the explosion starts to destroy the time machine, the laws we are using to predict the explosion – the laws of quantum fields in warped spacetime – break down. They fail, and they report to us that the remainder of the explosion is in the domain of quantum gravity. If we fully understood the laws of quantum gravity, we would just switch laws and keep calculating. But we don't, so we can't. And we won't know for sure the explosion's outcome until we have mastered the quantum-gravity laws.

We won't know till then, but I'm willing to speculate. Most of the hints from our calculations now point in one direction: it seems likely that *the explosion will always destroy the time machine.*

Carl was prescient in his novel. He was very careful to arrange that Eleanor Arroway return to Earth a fraction of a second after she left, not before. Somehow he knew that he should avoid backward time travel, but that he could come very close to it.

Thank you, Carl, for getting me involved in this. It's been a wonderfully fun game, but it has also been fruitful. Thanks to you we are beginning to ask new questions that will lead us to a deeper understanding of space, time, and the Universe.

BIBLIOGRAPHY

Flamm, L. 1916. Beitrage zur einsteinschen gravitationstheorie. *Physik Zeitschrift* 17:448–454.

Frolov, V. P. 1991. Vacuum polarization in a locally static multiply connected spacetime and a time machine problem. *Phys. Rev. D* 43:3878–3894.

Gödel, K. 1949. An example of a new type of solution of Einstein's field equations of gravitation. *Rev. Modern Phys.* 21:447–450.

Gott, J. R. 1991. Closed timelike curves produced by pairs of moving cosmic strings: exact solutions. *Phys. Rev. Letters* 66:1126–1129.

Grant, J. D. E. 1993. Cosmic strings and chronology protection. *Phys. Rev. D* 47:2388–2394.

Hawking, S. W. 1992. The chronology protection conjecture. *Phys. Rev. D* 46:603–611.

Hiscock, W. A., Konkowski, D. A. 1982. Quantum vacuum energy in Taub-NUT (Newman-Unti-Tamborino)-type cosmologies. *Phys. Rev. D* 6:1225–1230.

Kim, S.-W., Thorne, K. S. 1991. Do vacuum fluctuations prevent the creation of closed timelike curves? *Phys. Rev. D* 43:3929–3947.

Klinkhammer, G. 1991. Averaged energy conditions for free scalar fields in flat spacetime. *Phys. Rev. D* 43:2542–2548.

Morris, M. S., Thorne, K. S. 1988. Wormholes in spacetime and their use for interstellar travel: a tool for teaching general relativity. *Am. J. Phys.* 56:395–412.

Sagan, C. 1985. *Contact.* New York: Simon & Schuster.

Schwarzschild, K. 1916. Uber das gravitationsfeld eines massenpunktes nach der einsteinschen theorie. *Sitzungsberichte der Deutschen Akademie der Wissenschaften zu Berlin, Klasse fur Mathematik, Physik, und Technik.* 1916:189–196.

Thorne, K. S. 1993. Closed timelike curves. In R. J. Gleiser, C. N. Kozameh, O. M. Moreschi, eds., *General Relativity and Gravitation 1992: Proceedings of the 13th International Conference on General Relativity and Gravitation.* Institute of Physics Publishing: Bristol 295–315.

Thorne, K. S. 1995. *Black Holes and Time Warps: Einstein's Outrageous Legacy.* New York: W. W. Norton.

Wald, R. M., Yurtsever, U. 1991. General proof of the averaged null energy condition for a massless scalar field in two-dimensional curved spacetime. *Phys. Rev. D* 44:403–416.

INTERLUDE

This article was a Public Lecture delivered by Carl Sagan on October 13, 1994, at Cornell University, at the symposium honoring his sixtieth birthday. Dale Corson, Emeritus President of Cornell, introduced the lecture, which was followed by a question-and-answer period. Both the introduction and Sagan's answers to numerous questions contain interesting material for the reader. Thus, we reproduce here the proceedings of the entire evening.

Carl Sagan and Ann Druyan with President and Mrs. Frank Rhodes at the banquet concluding the symposium in honor of Carl's 60th birthday.

Carl Sagan with members of his close family at the symposium in honor of Carl's 60th birthday.

Carl Sagan with Yervant Terzian (left) and Frank Drake (right) at the symposium in honor of Carl's 60th birthday.

Carl Sagan and Ann Druyan in the early 1980s.

11
The Age of Exploration

CARL SAGAN

YERVANT TERZIAN: Good evening, ladies and gentlemen. It is my privilege to introduce to you Emeritus Professor of Physics and Emeritus President of Cornell University, Dale Corson, who will introduce Carl.

DALE CORSON: Professor Terzian, ladies and gentlemen, it is my pleasure to introduce Carl Sagan for his lecture, "The Age of Exploration." Dr. Sagan first crossed my consciousness one day in 1967, when Professor Thomas Gold, chairman of our Astronomy Department at the time, came to my office – I was university provost at the time – to tell me about Carl Sagan, a promising young astronomer at Harvard, whom Tom said he thought we could get. He told me that Sagan was a brilliant planetary scientist, and, furthermore, he had a great ability to tell the lay public in understandable terms what astronomy and science are all about. This latter characteristic, particularly, caught my attention. It represented a unique ability and we needed more of it in the university. I responded that this seemed like a good opportunity, and why didn't he make an offer? Professor Gold said he had no money, and that I would have to provide the funds. Now, this is a game that every entrepreneurial department chairman tries to play, but I didn't dismiss Gold as quickly as I might have some people. I'd known Tom for a long time. I'd been chairman of the search committee that brought him to Cornell, also from Harvard. Harvard is a good recruiting ground. Tom's always been an exciting person to have around, with more ideas per second than anyone else, and I've always enjoyed talking with him. His ideas about everything, the expanding universe, how to ride a unicycle, pulsars, Carl Sagan, but there's a problem. Sometimes his ideas are wrong. On the theory of riding the unicycle, for example, I think he never learned, but usually he's right. He was right about pulsars from the first moment. He told me that if I put up the money to hire

Carl Sagan, I would never regret it. I did put up the money; the offer was made, and Carl came to Cornell, and I have never regretted it. Tom was right.

You already know all the great things Carl has done in the past quarter century, although you may not appreciate all the solid science he has done. You can take my word for that. I've always been grateful to Carl for his willingness to talk to alumni groups and to other lay groups. When I was president, I asked him to do this billions of times, when I could find him, and he always said yes. He did hesitate once, when it was a black-tie affair in Chicago, but when I explained the importance of the occasion, he accepted. I think he rather liked the black-tie part of it, and I'm not sure he ever returned the rented tuxedo. Maybe that's why he's never been able to go back to Chicago.

Carl has received more honors and awards than I could possibly relate. Let me limit this reference to reading the citation for one of his recent honors, the Public Welfare Medal of the National Academy of Sciences, the academy's most prestigious award. "For his ability to communicate the wonder and importance of science, to capture the imagination of so many, and to explain the difficult concepts of science in understandable terms." That says it all. Carl Sagan, on "The Age of Exploration."

CARL SAGAN: Thank you, Dale. I never knew that Tommy hit you up for my salary. I'm grateful to you both. It's true that Tommy Gold recruited me for Cornell. I remember the inducements: a very small and exceptionally good astronomy department, superb ancillary departments in physics, chemistry, and biology, a beautiful campus, laboratory facilities which were by some standards very generous. But still, I hesitated. Tommy made the final inducement, I think knowing full well what he was doing. He took me on a little trip to Upper Enfield and I thought, my goodness, here is something as exquisite as any national park I've ever been in, and it would be right on my doorstep. That was the missing ingredient. Tommy was extremely persuasive on every level of inducement, and I thank him very much for the invitation. I've lived now in Ithaca longer than I have lived in any other place in my life, and I'm extremely grateful to Cornell and the town of Ithaca. I consider this my true roots.

We humans have enjoyed civilization only for about 10,000 years. Our species is a few hundred thousand years old. Our genus, the genus *Homo*, is a few million years old. Therefore, for the vast bulk of our tenure on Earth, we were something other than sedentary and – the word has such an aura of self-satisfaction – civilized. What were we? We were hunters and foragers. We wandered in small, itinerant, extended families. Our current knowledge of the hunter-gatherer life style is due mainly to a few courageous and far-seeing anthropologists who went and lived with the few remaining hunter-gatherer groups before they were finally and irreversibly destroyed by civilization. The anthropologist from whom I learned the most about hunter-gatherers is with us at this symposium, Richard Lee of the University of Toronto. He studied the !Kung bushmen of the Kalahari Desert in the Republic of Botswana. I want to give a little flavor about our ancestors from Lee's study of the !Kung.

It is very important to note that they are highly technological. The technology is wood and stone and domestication-of-fire technology, but it's unambiguously technology. They are technological because their lives depend upon it. Chipping and flaking stone tools back before the external civilization sent a little trickle of metal into their economy is key. They did it superbly well. The archaeological and anthropological record is clear that we were technologists all the way back to the beginning. So the idea that science and technology is something new, unusual, and inaccessible to most people is completely backwards. Technology is, if anything, the most characteristically human activity, although, as I'll mention later, it is not exclusively a human activity.

Now, hunter-gatherer game tracking techniques: A small group, with their bows and poison arrows and digging tools and a few other lightweight technological contrivances, is following the game. They come near a stand of trees. They take one close look at the ground. Immediately, they know how many animals went by, what their ages and sexes were, how long ago they passed; this one is lame in the back left foot; at the pace they're going we should be able to overtake them in another 2 hours if we hurry. Now, how do they know all this? In fact, what do they notice in order to follow the game on which their lives pretty well depend? One thing is the hoofprint. Different animals have different characteristic shapes of their hooves; different sized animals leave different sized hoofprints; but the decay of the hoof crater, the falling of pebbles in, the collapse of the raised rims, debris blown into it, tells you age. In fact, it reminds me of nothing so much as determining the ages of planetary surfaces by looking at how fresh the impact craters are. Maybe the reason that studying cratering physics seems so natural to us planetary scientists is because we've been doing it for a million years.

The !Kung also know that animals in the hot Sun like to avoid sunlight. If there is a shadow on the ground, they will deviate from their path to run through the cool shadow. But where the shadow is depends on where the Sun is, and therefore, when you see the deviation of the trail from a straight line, you know that there had to be a shadow at that spot when they passed. Well, where in the sky did the Sun have to be in order to cast that shadow? Oh, it was eleven o'clock this morning.

Now, I don't claim that every hunter-gatherer made such a scientific calculation, did the trigonometry of the angle of the Sun, and so on.

This was tradition; each generation taught the next. But someone had to have figured it out, and that someone had to be a scientist. This is another reminder that we've been scientists and technologists from the beginning.

Consider now the important and rueful fact that every human culture has considered itself at the center of the universe. What's this about? I think it's very straightforward. Back then, in hunter and forager times, many modes of modern nocturnal entertainment were unavailable. Television was not available, so over the dying embers of the campfire, people watched the stars. Why? One, it is straightforward dazzling. We today, living under chemically polluted skies, with nearly ubiquitous light pollution, have mainly forgotten how gorgeous the night sky can be. It is not only an aesthetic experience; it also elicits unbidden feelings of reverence and awe.

Secondly, people made up stories about the stars. They invented Rorschach tests up there, follow the dots, constellations. "Look like a bear to you, Og?" "Yes, I guess it does," and they forced their children to memorize these absolutely arbitrary patterns. Then myths were invented, either before or after, so these were visual reminders of events. "That's the bear that ate your grandpa," something like that, and gramps was put up in heaven as an example.

But beyond that, there was something enormously practical. The stars by their rising and setting constitute a great clock and calendar in the sky. In the absence of artificial timepieces, that's extremely important, because there are certain seasons of the year when the herds are running. There are certain seasons when the trees are ripe with nuts or fruit. If you know what those seasons are, and you know what the moment is, you can prepare, and you can eat.

Now, the most superficial examination of the sky shows the stars are rising in the east. Some of them pass directly overhead and some of them pass on small circles close to the horizon, but they all rise in the east, they all set in the west, and then in the daytime, they do something else. They somehow go around the bottom of the Earth that none of us has ever seen – it's flat as a board, of course – and then, the next morning they come up again in the east. Now, there's absolutely no doubt from this fact that the stars, the planets, the Sun, and the Moon, all go around us. We're obviously not moving. So we reside, immobile, at the center of the universe. An observed fact. Anybody who denies it must have something wrong with him. This is the geocentric conceit.

Now, not only did every culture draw this conclusion, but our ancestors took enormous personal satisfaction in it. Think about it: we are at the center of the universe. The center of the universe is surely an important place. Not only that, what other animals, what plants, make use of the apparent motion of the stars? Only us. Therefore, the stars have been put there for our benefit. The Sun and the Moon also have obvious practical uses. There was some confusion, though. Maybe you know the old story about the Persian wise man and philosopher who was asked, "Which is more useful, the Sun or the Moon?" and replied, "Of course, the Moon, because the Sun shines in the daytime when it's light anyway, whereas the Moon only shines at night, when we need light." The centrality of our position was stunning.

I imagine an extraterrestrial visitation of the sort that there is absolutely no evidence for, coming upon the Earth, which is running around the Sun once every year and then listening in on what people all over the planet are saying: "We're at the center! We're important! We're special! Everything goes around us!" And then I imagine the extraterrestrials thinking of the Earth as the planet of the idiots. But that's too harsh, because there is a resonance here between the most obvious interpretation of absolutely straightforward observational facts that every person can verify for him or herself, a resonance between that and our emotional hopes and needs. The idea gained currency that the universe is made for us, not because of any particular merit of ours, but just because we're here or just because we're human. To me, this seems to resonate with the same psychic wellsprings responsible for the view that our *nation* is special and the center of the universe (which, by the way, is the literal meaning of The Middle Kingdom for centuries applied by the Chinese to China). The names that countless ethnic groups apply to themselves – !Kung, Hopi, Alimani – translate essentially as The People. *We're* people. Those other guys, they're something less than people. The same psychological roots that convince us of the superiority of our gender, or our ethnic group, or the particular melanin content in our skin, or our particular language or headdress or clothing styles or convention of pulling out the handkerchief when we sneeze, or almost anything else. We have a weakness for chauvinism. Scientists are creatures of the culture in which they swim, and so we also are vulnerable to this chauvinist or geocentrist or anthropocentrist conceit.

Except for a nearly invisible blip associated with the name of Aristarchus of Samos, we went on, every human culture, every great philosopher, every scientist, every religious leader, thinking we were at the center of the universe. We put it in various guises in our sacred scriptures, and declared the scriptures to be infallible, thereby making it not just a secular but a religious crime to question the conclusion.

In the late fifteenth century, an astronomer-cleric from Poland named Nicholas Copernicus thought he had an alternative idea, namely that the Sun was at the center, and the Earth, like the other planets, went around it. He knew that this was dangerous stuff, and so withheld the publication of his book until he was on his deathbed. Even then, the way it worked out, when the book was published, it had a preface written by a well-meaning friend of his, Andreas Osiander, which essentially said (I'm paraphrasing), "Dear reader, when you look at this book, it may appear that the author is saying that the Earth is not at the center of the universe. He doesn't really believe that. You see, this book is for mathematicians. If you wish to know where Jupiter will be two years from next Wednesday, you can get an accurate answer by assuming that the Sun is at the center. But this is a mere mathematical fiction. It does not challenge our holy faith. Please have no anxiety in reading this book."

This peculiar split-brain compromise actually lasted for almost two centuries – and how many other tawdry compromises between conventional wisdom and new ideas we've bought into since then! Finally, Galileo made a forthright and brilliant defense of Copernicus, based in part on a set of observations from the newly invented astronomical telescope, and the Church grew increasingly annoyed. Galileo remained obdurate. I once had the pleasure at the behest of Franco Pacini, then director of the Arcetri Observatory in Florence, who is also with us here today, to actually trod in Galileo's footsteps and hold a close replica of his tiny telescope. It was so modest to have worked such a revolution. When Galileo became too insistent, the princes of the Church exhibited to him the instruments of torture in the dungeons of the Inquisition. (They weren't making any particular point; they just thought he'd like to broaden his general perspective.) Shortly thereafter, Galileo made his famous confession in which he abjured the abominable doctrine that the Sun and not the Earth was at the center.

But the stage had been set. Truth will out. The debates went on, and when in the eighteenth century Bradley discovered the aberration of light, and then in the nineteenth century the long-sought stellar annual parallax was found, the opposition collapsed. Now, everybody is taught that the Earth is not at the center of the universe. Except I think there's much evidence that we are all still covert geocentrists, with a heliocentric veneer. Think, for example, about our language. Sunrise. I was up before sunrise. Sunset. It was a gorgeous sunset. But the Sun isn't rising or setting. The Earth is *turning*. Think of how difficult it is for us to simply parse a simple word or phrase that conforms to the Copernican perspective. "Billy, be sure to be home before the rotation of the Earth makes the local horizon occult the Sun." Billy is gone before you're half-way through. Why isn't there any snappy phrase like sunrise or sunset in the Copernican context?

Recent opinion polls show that 25–50% of adult Americans do not know that the Earth goes around the Sun and takes a year to do it. In China, the figure is 70%. Bear in mind that the Copernican perspective gets a lot of press in the United States, and still at least a quarter of us have missed it, and that in China, where there isn't a NASA, and where the television programs are much less sophisticated, a much larger percentage of people have missed it. If anything like China is typical, it may be that today, five centuries after Copernicus, most people on this planet still think in their heart of hearts that the Earth is at the center. So, congratulations on our wisdom in deducing our true cosmic circumstances may be premature. (At the same time, that any of us figured any of this out is, I think, a legitimate cause for pride.)

In my view, a great deal of the history of science can be understood in terms of something like this Copernican debate. In many cases, the going-in position is that we're central or important, there's something fantastic and great about human beings. There is then actual observation of our circumstances; nobody ever thought to look before. And then, the result is the daunting and disquieting discovery: no, we're not at the center; no, we're not important. To my mind, many of the key findings of science, much of the modern scientific perspective, evolves from debates with just that character – what Ann Druyan has called The Great Demotions.

Shortly after Copernicus, there were people who said, "OK, maybe we're not at the center of the universe; maybe the Sun is; but we're close to the Sun, so we're almost at the center of the universe; it's almost as good." *Was* the Sun at the center of the universe, which we can loosely translate as at the center of the Milky Way galaxy? No. We are not at the center, where it does look important, or at least well-lit. Instead, we are near an obscure spiral arm 30,000 light years from the center of the Milky Way galaxy, in the galactic boondocks. If you were an intergalactic traveler approaching the Milky Way, what would you think of someone who said, "Excuse me, captain. That's the center of the Milky Way, it's true, a galaxy of four hundred billion stars. But count out spiral arms with me. See, there's one, there's another one, really big and beautiful. There's another one, then over there, you see that obscure spiral arm? Well, don't look exactly in it, but just a little beyond it. See over there? I know it's hard to see. Take a closer look at the stars here. No, not that one. That one, see? The beings who live on that one say they're at the center of the universe, and that the entire universe is made for their benefit." What would you think of those guys? And then suppose it emerged that not one soul on that planet thinks otherwise. Every one of them is convinced they are at the center of the universe, and the reason there is a universe.

Then, it was thought that at least the Milky Way is the only galaxy. But no, not only are there other galaxies, there may be as many as a hundred billion of them. There was a moment when the Hubble Flow was discovered, when it was found that the galaxies are all running away from us, the more distant galaxies running away the faster. There were people who breathed a sigh of relief. This was in the twenties of this century. We're not at the center of our galaxy, but, it seemed, our galaxy is at the center of the entire universe. This is based upon a serious misapprehension. There is no center to the universe, at least in ordinary three-dimensional space, and astronomers in any one of these galaxies would see all the other galaxies running away from them in the same way that we do.

For a long time, all through my growing up and undergraduate and graduate school career, astronomers believed that there were no other planetary systems. And if there were no other planets, while life has to arise on planets, then there's no other life, no other intelligence, and so in this sense, at least, we're at the center of the universe. One of the hallmarks of the age we live in is that this chauvinism too is in the process of teetering and collapsing. We find that more than half the nearby young sunlike stars have circumstellar discs of gas and dust, extremely like what has been deduced for the circumstances from which the planets formed in our solar system. The key datum is that the orbits of the planets are very largely coplanar. (This was, by the way, something that Isaac Newton, no less, thought he could deduce the hand of God from. That is, God established initial conditions for all the planets in the same plane. But Kant and Laplace showed that the conservation of angular momentum meant that an irregular spinning, contracting cloud would collapse into a thin disc and that planetary formation would occur in the disc, guaranteeing coplanarity.)

Not only are such circumstellar discs amazingly numerous, but we now have the first bona fide extrasolar planetary system going around a star that must be at the bottom of the list of potential candidates that anyone would have imagined, a pulsar named B1257+12. This particular pulsar is something like an atomic nucleus the size of the Cornell campus, spinning at 10,000 rpm – 10,000 revolutions per minute. It's a supernova remnant. There was a colossal catastrophe that blew off most of the mass of that star. Going around it are at least three planets, two roughly of earthlike mass, one roughly of lunar mass, a little in closer than the Earth, Mercury, and Venus. [Since this symposium, three planets of roughly Jovian mass have been discovered orbiting around the stars 51 Pegasi, 47 Ursae Majoris, and 70 Virginis.] The processes that lead to planets look to be very broad and general. The technology is now improving, so that in the next ten, twenty, thirty years, in the lifetime of most of the students in this audience, we ought to have completed comprehensive surveys of the nearest few hundred stars, maybe much more than that, to detail what planetary systems they have. I think we can be fairly confident of other Earths among them.

In short, we live on a hunk of rock and metal that orbits a humdrum star in the obscure outskirts of an ordinary galaxy comprised of 400 billion stars in a universe of some hundred billion galaxies, which may be one of a very large number, perhaps an infinite number of separate, closed-off universes. Many, perhaps most, of those stars probably have planets. In this perspective, how can anyone seriously believe that we are central – physically, much less to the purpose of the Universe?

There has been the view that if there is nothing special about us in space, maybe there is something special about us in time. We've been put here by the Creator to take care of things. Stewardship is the engaging word often used. Who knows what dire consequences would happen to the environment without us? So, we have an obligation to make sure everything goes as God would have wished it. To me, the principal trouble with this idea is that 99.998% of the lifetime of the universe from its beginning to now was over before any human appeared on the scene. So if we are the caretakers, where have we been for most of the time we were supposed to be doing our job? We have been terribly lax. I could see that the Chief Gardener might be very annoyed with us, which in turn might explain a great deal.

If there is nothing special about our position in space or our position in time, maybe there is something special about our motion. We have a privileged frame of reference: This was the classical absolute motion physics that every great physicist bought into until 1905. Albert Einstein, all his life a keen critic of privilege in the social sphere, immediately mistrusted the contention that the planet we happen to live on was affixed to an imaginary frame of reference that had special merit with regard to the laws of nature. Instead he asked, what kind of physics would be implied if you deduced the same laws of physics no matter what planetary system you lived on. Out of this question flowed special relativity, which is repeatedly confirmed by experiment.

The notion of centrality and unmerited importance has been even more pervasive in the biological than in the physical sciences. It is particularly stark in the conceit called special creation. More than any other creatures, it asserts, the Creator of the universe has plans for us, key to the reason there *is* a creation. We're different from the other animals, not just in degree but in kind. No one else has morals, ethics, altruism, compassion, courage; no one else can foresee the future consequences of present actions; no one else has art and music; nobody else can use tools; no one else can make tools. This list, it goes on and on, is essentially agreed to by Plato, Aristotle, Augustine, Aquinas, Hobbes, Locke, all the great figures in philosophy, with the single exception of David Hume (hats off to him), bought into by all scientists, including highly skeptical ones, up until the year 1859, and enthusiastically adopted, of course, by all the religious leaders of the Judaeo–Christian–Islamic religion.

In 1859, Charles Darwin made the first and heroic effort at deflating this conceit, showing that one species could evolve by natural processes, without anything foreordained, from another. More than a decade later, he published his second book on the subject, suggesting that evolution applies explicitly to us. He proposed that all of us Earth beings are relatives, with a single common ancestor. The implication that we arose through the laws of chemistry, perhaps out of primeval slime, was unsettling – to many, repulsive. It stood in stark contrast to the uplifting idea, however bereft of explanatory power it may be, of the hand of a loving God.

We and the chimps have a most recent common ancestor who lived a few million years ago. Chimps are our cousins. Even today, this contention upsets many. "Have you been to a zoo?" they ask. "Have you looked at what a chimpanzee does? Maybe you are related to chimps, buddy, but I'm not!" Well, we can learn about chimp behavior in zoos about as well as we can learn about human behavior in jails, and for exactly the same reasons. Prison doesn't bring out the best in us. When people like Jane Goodall devote themselves to observing chimpanzees in their natural habitats, the chimps get used to them. Then we find very different behavior.

I can't resist telling the story of Geza Teleki, an anthropologist and student of animal behavior who wished to learn chimpanzee technology, particularly the termite-fishing industry in which they are adept. So he apprenticed himself for 9 months to a chimp named Leakey, an expert. Here's how the chimpanzee termite-fishing industry works: You find a reed, not any kind of reed, but the right kind of reed. You strip it of supernumerary branches, check that it has the right heft and suppleness, and then approach the enormous termite mound, an apartment house for termites. Each night the termites cover over the entrances to their nests. The chimp takes a look, scratches away two or three places where the entryways have been walled up, takes the reed or grass stem, in one deft motion inserts it into the termite mound, gives a few twists, carefully pulls it out, and the tool is covered with termites. The chimp goes, yum, yum, yum, yum and has provided herself an excellent source of protein. If a human were to be dropped down, on his own, in this same place in East Africa, with a need for protein, how would he do? Teleki, spending full time on this problem for months, could not break off the right kind of reed; in fact, he had to use the reeds that chimps had discarded. He could not after 9 months find the walled-up openings to the apertures that Leakey could identify at a single glance. Teleki could not put the reed in deftly (for example, the reed would come out accordioned). Teleki could not wiggle the reed enticingly to get the termites on, and could not withdraw it without scraping off almost all the termites. At the end of 9 months, he had failed utterly.

There is a bonobo, a so-called pygmy chimp, named Kanzi, who lives in Atlanta, who not only knows how to use stone tools but also knows how to make stone tools. Chimps know how to do things. How do they know when a human with a Ph.D. does not? They were apprenticed for years. In the acknowledgments to Teleki's wonderful paper, he thanks his patient tutor, and apologizes for his own failures, because "they are not the fault of the being he was apprenticed to." It's just that humans are not very good at learning some things.

Now, of course, there are differences between chimps and us. We have electric light bulbs, police cars, compact disc players, and nuclear weapons. Chimps don't. But we can't say that they don't have *any* technology, and for most of human history we didn't have these technologies either: Stone tools and fire were the technological frontiers.

When it became possible in the late twentieth century to do DNA-base sequencing, you could get a quantitative measure of the relationship between humans and chimps, and it turns out the two species share 99.6% of their active genes. One way to look at it, I guess, is to conclude that 0.4% is a much larger number than we had guessed. Another way to look at it is, if you want to know about us, study chimps. There's a lot there to learn. In any case, the idea of Special Creation is really an idea from another time. If not much else but the molecular biological evidence were available, it would still be very clear that there is nothing in us that is qualitatively different from our nearest chimpanzee relatives – with the possible exception of a talent for language.

Where are the battles on the Great Demotions taking place today? One is on the search for extraterrestrial intelligence. We have not found life, much less intelligence, anywhere else yet. We send spacecraft to other planets to look for microbes. We construct large radio telescopes and listen to hear if anyone is sending us a message lately. Both of these activities have led to occasional tantalizing data, but none of it of sufficient quality and repeatability to say that we have detected life or intelligence elsewhere. In our ignorance, the geocentrists find hope. They confuse absence of evidence with evidence of absence. "You haven't found life elsewhere? There isn't any! Those who live on Earth are the only living creatures in the universe. You haven't found intelligence elsewhere? There isn't any. It's only here. We are at the center of the intellectual cosmos." While I could give you what I consider to be a strong plausibility argument why this conceit is also erroneous, it is only fair to say that nobody knows the answer to this question. We have not yet found life or intelligence elsewhere. We are in the course of looking. Maybe we will find it tomorrow, or maybe it will take centuries. Maybe we will never find it. All we have to do is keep an open mind. There is no other approach. You don't have to make up your mind in the absence of evidence.

Finally, there is a new, and to my mind, bizarre field for this debate, something called the anthropic principle. It would with greater justice be called the anthropocentric principle. It comes in strong, weak, and various shades of middling flavors. The weak anthropic principle asserts that if the laws of nature and the fundamental constants of nature were significantly different, then the paths that in fact led to us would have been altered, and we wouldn't be here. That is unexceptionable; it's certainly true. But the strong anthropic principle is to my mind dangerously close to the following argument: we would not be here if the laws of nature and the values of the physical constants were other than they are; therefore, the laws of nature and the physical constants are as they are in order for the universe to produce us. God had us in mind at the time the universe was made, and here we are – back at the center of the universe at last.

There are many things that can be said about this: Who has traced through what other laws of nature and physical constants are possible and mutually consistent, but also lead to the functional equivalence of life and intelligence? You can also note that the strong anthropic principle is not very amenable to experimental investigation. There's something telling about calling it the anthropic principle, because many of the same laws of nature and the same physical constants are required to make a rock as to make a person. Why is it not called the lithic principle, with strong and weak versions, and in the strong lithic principle, the laws of nature and the physical constants are as they are so rocks could come into being? If rocks could philosophize, I bet we would hear nothing of the anthropic principle. At the cutting edge of rock philosophy would be the lithic principle.

* * *

I want to conclude with one of the many psychic rewards that planetary exploration has brought to me. As Ed Stone outlined, there was a moment when the two Voyager spacecraft had completed their closeup reconnaissances of the Jupiter, Saturn, Uranus, and Neptune systems. The spacecraft phenomenally outperformed their design specifications. The bulk of our knowledge of the outer solar system has come because JPL did such a brilliant job with these extraordinary spacecraft – coming in on time, under cost, and vastly exceeding the fondest hopes of their designers. There were no other planets that we were going to rendezvous with further out in the solar system. It was now possible to turn the cameras close to the Sun and if the worst happened and we burnt out the optics, so what? There was nothing else we were going to photograph. I had wanted, from the time of the Saturn encounter, to take a picture of the Earth from the most remote vantage point. But it was by no means easy, even though the downside was almost nil, to turn the cameras back, and it required an actual intervention by the NASA administrator to get it done.

It was clear that in such a picture, the Earth would appear only as a single picture element, a pixel. You would not even see continents. You could not tell any detail. I still thought it would be useful to do in the same sense that the great frame-filling Apollo 17 picture of the whole Earth has become a kind of icon of our age – because it said something very powerful to us, including the fact that from that perspective, national boundaries were not in evidence. Here it is: the Earth from Voyager 1, momentarily in a sunbeam (Plate X). Take a look. From the outskirts of the planetary part of the solar system, it's a pale blue dot.

That's us. That's home. That's where we are. On it, everybody you loved, everybody you know, everybody you have ever heard of, lived out their days. The aggregate of all our joy and suffering, thousands of confident ideologies, religions, economic doctrines, every hunter and forager, every hero and coward, every creator and destroyer of civilizations, every king and peasant, every young couple in love, every hopeful child, every mother and father, every inventor and explorer, every revered teacher of morals, every corrupt politician, every superstar, every supreme leader, every saint and sinner in the history of our species, lived there.

The Earth is a very small stage in a great cosmic arena. Think of the rivers of blood spilled by all those generals and emperors, presidents and prime ministers and party leaders, so that in glory and triumph they could become the momentary masters of the corner of a dot. Think of the endless cruelties visited by the inhabitants of one part of the dot on the scarcely distinguishable inhabitants of another part of the dot. How frequent their misunderstandings. How eager they are to kill one another. How fervent their hatreds. Our posturings, our imagined self-importance, the delusion that we have some privileged position in the universe, seem to me challenged by this point of pale light. Our planet is a lonely speck in the great enveloping cosmic dark. In our obscurity, in all this vastness, there is no hint that there is anyone who will come and save us from ourselves.

It's been said that astronomy is a humbling and, I would add, character-building experience. To me, this is one of many demonstrations, through astronomy, of the folly of human conceits. To me, this picture underscores our responsibility to deal more kindly with one another and to preserve and cherish the pale blue dot, the only home we've ever known.

If we have an indomitable urge to be central and important, we know what we can do. We can take seriously the admonition to compassion and loving kindness of the major religions that most of us profess to live by. We can work toward the economic self-sufficiency of the billion poorest people on the planet. We can make sure we know what dangerous predispositions lie within us and what desperate mistakes we have made in our history. We can make it unthinkable for our children to grow up ignorant. We can care for and cherish the planetary environment that sustains us and the other beings that share it with us. We can, every one of us, take a fierce and principled role in the democratic process. We can insist on the honesty and selflessness of our elected officials. And we can judge our progress by the courage of our questions and the depth of our answers, our willingness to embrace what is true rather than what feels good. If we want to be important – not passively because of the species or ethnic group we have by accident been born into, or the planet we have been born on, but because of the merit of our actions – we know what we can do.[1]

CORSON: Thank you, Carl, for a great speech. I'm already looking forward to your seventieth birthday. Professor Sagan will answer questions for a while.

Q: Please, we'd like to know a few things. The first one is, who said, billions and billions?

SAGAN: It's someone named Johnny Carson. I never said it. (Of course, since *Cosmos* I've said it a few times answering questions like this.) It's sort of like Humphrey Bogart never said, "Play it again, Sam." Nobody believes it, but he ...

Q: We also want to know, what, if any one thing in particular, would you have yourself be remembered by?

SAGAN: I have to leave the decision about how I'm going to be remembered, which I hope will not have to be faced for some time, to others. There's still a great deal I hope to do. But thanks very much for the question.

Q: I was interested in your great demotions, and I certainly agree in most ways with what you say. Where, among the great demotions, would you fit in, or would you fit in at all, the appearance of consciousness among humans?

SAGAN: Yes, that's an important question.

Consciousness has various meanings. If it means an awareness of the external world, and modifying your behavior to take account of the external world, then I think microbes are conscious. How do you know that I think any thoughts? Only because I happen to be communicating to you. You can't easily tell that I have philosophical thoughts by looking at me drinking this cup of water, right? So imagine that I was mute, that I could not communicate by speech or writing or anything else. Then how would you know if I had such thoughts? The evidence for not just the so-called higher apes, but running through all the apes

---

[1] Further details, and references, for Dr. Sagan's presentation can be found in *Shadows of Forgotten Ancestors: A Search for Who We Are* (with Ann Druyan) (New York: Random House, 1992; London: Century, 1992), *Pale Blue Dot: A Vision of the Human Future in Space* (New York: Random House, 1994; London: Headline Book Publishing, 1995); and *The Demon-Haunted World: Science as a Candle in the Dark* (New York: Random House, 1996; London: Headline Book Publishing, 1996).

PLATE I
A composite of planetary images returned from U.S. missions. Clockwise from top: Mercury, Venus, Earth and its Moon, Mars, Jupiter, Saturn, Uranus, and Neptune. (Plates I–V are courtesy NASA/JPL)

PLATE II

The Magellan spacecraft's imaging radar system, peering through the dense clouds of Venus, acquired data that were the basis of this global mosaic of the planet.

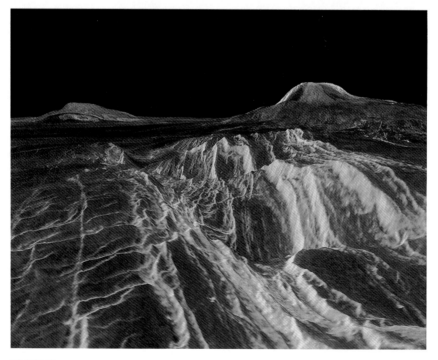

PLATE III

The vertical relief in this image has been exaggerated to highlight the large rift zones and volcanoes marking Venus' surface.

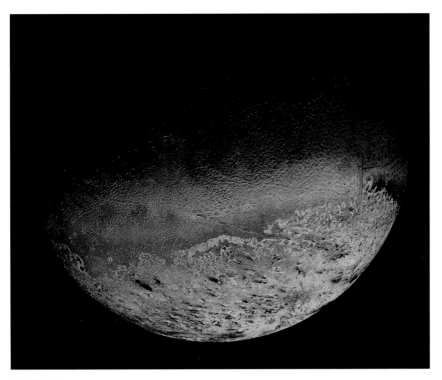

PLATE IV
The surface of Triton, one of the Neptunian moons, exhibits a brownish color suggestive of an organic residue. (Image processing by U.S. Geological Survey, Flagstaff, AZ)

PLATE V
Made up of organic smog particles, Titan's thick haze layer appears orange in this false-color image and lies below the thinner, blue layers of haze.

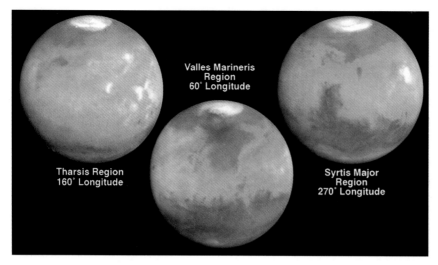

Valles Marineris
Region
60° Longitude

Tharsis Region
160° Longitude

Syrtis Major
Region
270° Longitude

PLATE VI
A telescopic view of Mars is not nearly as good as these Hubble Space Telescope images.

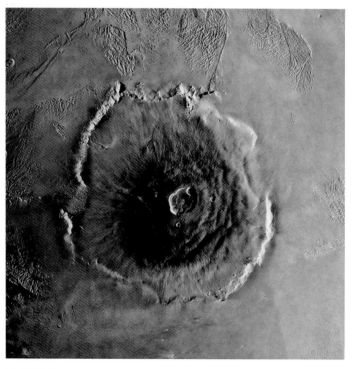

PLATE VII
This exquisite, highly processed picture of Olympus Mons, the largest volcanic structure in the solar system, vividly demonstrates how deceptive the early views of Mars were in the 1960s. In this scene, the sun is coming from the lower right. There is a large complex caldera in the center of the image that itself is large enough to include the whole island of Hawaii. At the base of the volcanic construct is a high escarpment, testifying to an ancient period of erosion after the main volcanic construct was emplaced. This structure is about 500 kilometers (350 miles) in overall dimension and rises nearly 29 kilometers (about 90,000 feet) above the mean Martian surface.

PLATE VIII

Impact of Comet Shoemaker – Levy 9 with Jupiter as viewed from one of the trailing fragments in the comet string. In this painting by Don Davis, one of the comet nuclei has just struck the planet, creating a bright flash as it explodes in the atmosphere; the fragment in the foreground will strike the planet in about 36 hours. The energy released in each impact is several million megatons, depending on the size of the fragment. (Courtesy NASA Ames Research Center)

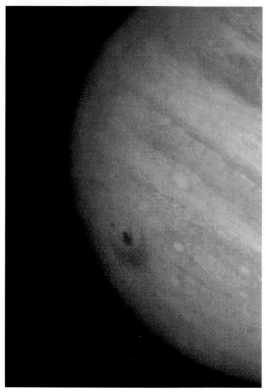

PLATE IX

The battered planet Jupiter as imaged by the Hubble Space Telescope on July 17, 1994, about two hours after the impact of fragment G of Comet Shoemaker – Levy 9. The sharp circular ring around the impact site is an expanding atmospheric wave caused by the explosion, whereas the asymmetric dark apron consists of ejecta from the explosion that has fallen back into the jovian stratosphere. The long-lived dark debris clouds from these explosions are comparable in size to the entire planet Earth. (Courtesy HST comet team and NASA)

PLATE X
Voyager 1 photograph taken from the outer edge of the solar system, showing the Earth as
a pale blue dot in a sunbeam.

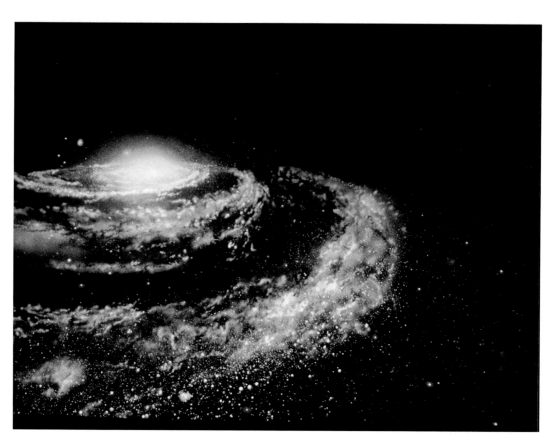

PLATE XI
"Approaching the
Milky Way Galaxy"
from *Cosmos*,
painting by Jon
Lomberg©.

PLATE XII
"Dinosaur's Last
Sunset" by
Jon Lomberg©.

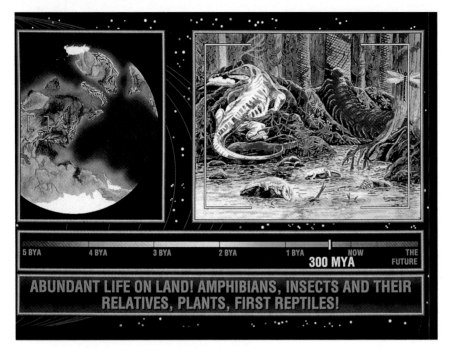

PLATE XIII

Image from a slide set on evolution produced by the Life in the Universe curriculum project of the SETI Institute in Mountain View, California. Slide art and design by Jon Lomberg and Simon Bell. ©LITU Project, SETI Institute.

PLATE XIV

"Portrait of the Milky Way" by Jon Lomberg. The most accurate image of our galaxy yet made. ©Artist and National Air and Space Museum.

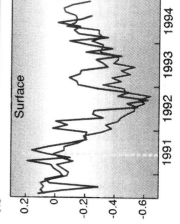

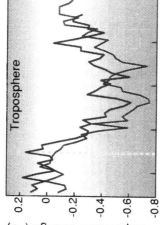

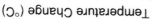
Temperature Change (°C)

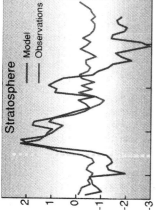

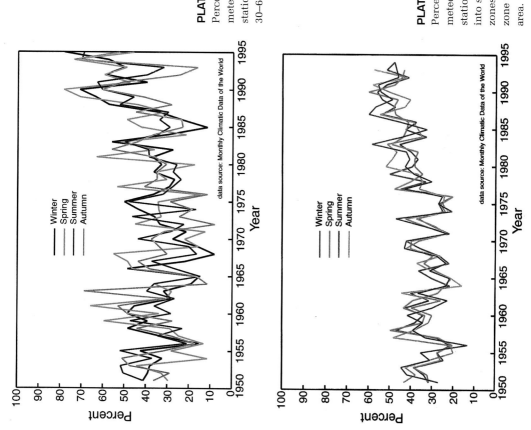

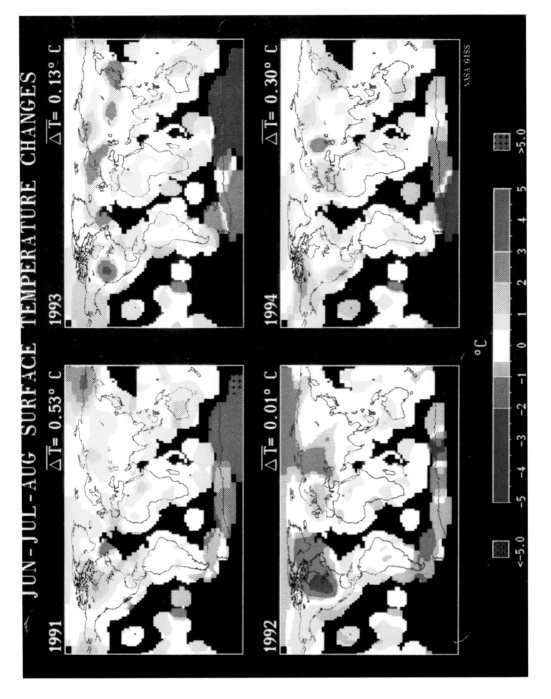

JUN–JUL–AUG SURFACE TEMPERATURE CHANGES

1991 $\overline{\Delta T}$= 0.53° C

1992 $\overline{\Delta T}$= 0.01° C

1993 $\overline{\Delta T}$= 0.13° C

1994 $\overline{\Delta T}$= 0.30° C

NASA GISS

°C

<-5.0   -5   -4   -3   -2   -1   0   1   2   3   4   5   >5.0

PLATE XVIII
June–July–August
surface air
temperature
anomalies, relative
to 1951–80 mean.

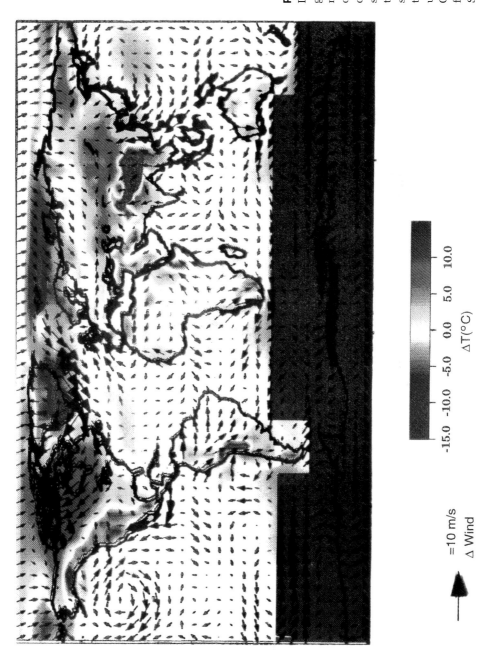

PLATE XIX
Difference between global climate model and observed climatologies for surface air temperature and surface winds, for the climate model used in the 1994 Goddard Institute for Space Studies Summer Institute.

ΔT(°C)

-15.0 -10.0 -5.0 0.0 5.0 10.0

=10 m/s
Δ Wind

Observations (1979-94)

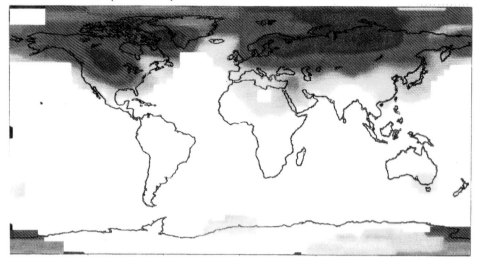

Model (No Forcing)

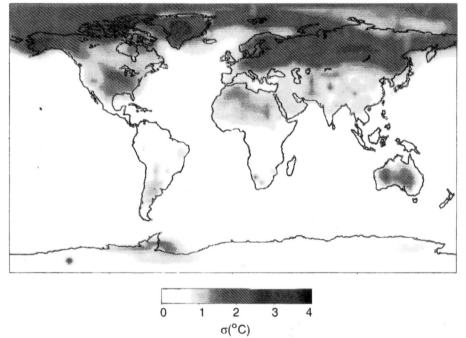

0   1   2   3   4
σ(°C)

PLATE XX

Observed standard deviation of surface air temperature for 1979–94 and the same quantity produced by the global climate model in the absence of climate forcings.

and monkeys, to me is very persuasive that they have thoughts, useful, practical thoughts, like lying, deceit, planning to fool others, thinking about it far in advance. Let me just give one little image which I like, because it covers many different grounds. These are the results of work in the Arnhem colony in the Netherlands, where there is a large, free-roaming community of chimps. Males are testosterone-driven and subject to raging hormonal imbalances. They get angry at each other and pick up rocks. They go quite a distance to get the rocks in order to confront the guy they're angry at, and throw stones at him. The very act of going out of sight of the enemy to collect stones and then bring them back to throw indicates thinking ahead, understanding a goal, and awareness of others. But the most interesting thing is, it's common for female chimps, seeing the males burdened with their stones, to walk up to them, open up their fingers, throw the stones away, and disarm them. When the males, in a huff, gather them up again, the females disarm them again. So, not only do the males have something in mind; the females know what they have in mind. That to me not only is consciousness, but a social arrangement I'd like to see more of in humans.

Q: My question is, given all these demotions, what is your personal religion, or is there any type of god to you; like, is there a purpose, given that we're just sitting on this speck in the middle of this sea of stars?

SAGAN: I don't want to duck any questions, and I'm not going to duck this one. But let me ask you first, what do you mean when you use the word god?

Q: Well, I guess my question is, it's like, is there a purpose for, I mean, given all these demotions, why don't we just blow ourselves up?

SAGAN: Let me turn the question around. If we do blow ourselves up, does that disprove the existence of God?

Q: No, I guess not. I guess what I'm asking is, since as we kind of make God almost go away in this, through these demotions. . . Through the ages, we have, humans have created a mythological framework that has always involved some kind of, often involved some kind of higher spiritual . . . If that goes away, as we know more and more, and it seems harder and harder to prove that anything might exist like that, where does that leave us?

SAGAN: On our own. [Applause] – which to my mind is much more responsible than hoping that someone from the outside will come and save us from ourselves – in which case we don't have to make our best efforts to do it ourselves. If I'm wrong, and there is someone who steps in and saves us, that's all right.

The word God covers an enormous range of different ideas (and you recognized that in the way you phrased the question) – running from an outsized, light-skinned male, with a long white beard sitting on a throne in the sky and counting the fall of every sparrow, for which there is no evidence, to the kind of god that Einstein or Spinoza talked about, which is very close to the sum total of the laws of Nature. Now, it's an observable fact, and a magnificent one, that there are laws of Nature, that apply through the whole universe. If that's what you want to call God, then of course God exists.

There are some other nuances. There is, for example, the deist god that many of the founding fathers of this country believed in, a *roi faineant*, a do-nothing king, the god who creates the universe and then retires, and to whom praying to is sort of pointless. He's not here now; he went somewhere else; he had other things to do. Now, that's also a god, a creator god, but very different from the Judaeo–Christian–Islamic one. So, when you ask, do you believe in God, if I say yes or if I say no, you have learned absolutely nothing.

Q: I guess I'm asking you to define yours, if you have one.

SAGAN: But why would we use a word so ambiguous, that means so many different things?

Q: It gives you freedom to define it.

SAGAN: It gives you freedom to seem to agree with someone else with whom in fact you do not agree. It covers over differences; it makes for social lubrication, but it is not an aid to truth. I think we need much sharper language when we ask these questions. Sorry to take so long in answering this, but this is an important issue.

Q: Me and some of the other students here at Cornell were wondering about your house. Was it some kind of power plant, or what was it before it was your house?

SAGAN: It's a study that Annie and I work in. It was, a long time ago, the headquarters of the Cornell Sphinx Head Society, in which God knows what abominable rites were performed! But I can assure you these days it is extremely placid. It also was once a sculpture studio of a remarkable Cornell professor, designer of nuclear accelerators and sculptor par excellence, Bob Wilson.

Q: I wanted to know what your views were on astrology; I know it's a related science to astronomy.

SAGAN: Astrology is a hoax.

Q: Excuse me?

SAGAN: It's a hoax. H-O-A-X.

Q: OK, what about documented proof of studies where extrasensory perception (ESP), clairvoyancy, . . .

SAGAN: That's different, right? Astrology is different from ESP and clairvoyance.

Q: OK, so I confused the two. . .

SAGAN: I'm happy to answer. I just wanted to be clear.

Q: OK, extrasensory perception, telekinesis; do they exist and do certain individuals have it? How did it arrive?

SAGAN: If it exists, it would have arrived by evolution, by natural selection, the same way as everything else. But what do we mean by extrasensory perception? There's an African fresh-water fish that establishes static electric fields and then detects its prey by perturbations in the electric field. We can't do that at all. It doesn't correspond to any of our senses. Does this fish have ESP?

Q: In a sense, yes, but, OK. . .

SAGAN: If it does have ESP, is this mysterious? Is it a challenge to science? Or is it just another way of perceiving the world?

Q: It's a different way of perceiving the world.

SAGAN: Yes. Therefore, if there is ESP, I think the chances are excellent that it can be well understood by science. But to the best of my knowledge, there isn't any ESP.

Q: Thank you. That answered my question.

Q: I'd like, first of all, to say that it must be a true privilege to be able to develop a career in something as stimulating, intellectually and spiritually, as astronomy. Now, my question to you is, what's your opinion on the use of animals in biomedical research?

SAGAN: I have struggled greatly with this issue, in part because I have a graduate student, Peter Wilson, who holds my feet to the fire. For example, I have a twenty-year-old leather jacket that I used to wear to Cornell that I don't wear anymore. I'm very conflicted on this issue. That gratuitous pain should not be inflicted on other animals is clear. That animals should not be made to suffer for fairly trivial goals, the making of lipstick, for example, I think is also clear. To argue, though, that animals should not be used in the pursuit of medicines and medical procedures that might save the lives of humans, is not so clear. Charles Darwin, far ahead of his time, made just the same distinction. If I had to explain, if somehow it was my job to do so, to people whose child was dying because a medical procedure was unavailable which might very well have been available if animal experimentation had been performed, I don't know how I would do that justification. Now, you might say to me that I'm attributing a higher value to humans than to other animals, and where do I come off doing that, especially at the end of an evening where I have been decrying chauvinisms? This to me has some resonance with the argument, why should we take any steps to save ourselves if an asteroid is going to hit the Earth? Asteroids have hit the Earth in the past and worked the extinction of other species. So why shouldn't we go quietly and the raccoons will have their chance, or the ants, or the sulfur-oxidation-state-altering submarine worms? At this point, I have no difficulty – since I happen to be, by an accident of birth, a human being – in justifying human beings trying to survive under sometimes trying circumstances. That's my judgment; I'm sure if I were a lizard, I would be arguing about sacrificing the humans so we can get better medicine for the lizards. I'm sorry. I can't help it. I'm a human.

Q: A Happy Birthday, Professor Sagan. I'm sorry I didn't get you a present. Now, there are various structures in the Andes and also formations in crops in northern England which people say are results of extraterrestrial appearances.

SAGAN: The plains of Nazca.

Q: I was just wondering about your views on whether or not we actually have been landed upon.

SAGAN: One way to look at it is this: Where did all this stuff about the plains of Nazca being mysterious and extraterrestrial come from? It came from a Swiss hotelier, Erik von Dänniken, who wrote a book called *Chariots of the Gods*, which became a worldwide best seller. He argued as follows: On the plains of Nazca in Peru there are large drawings. Some of them look like spiders, some look like turkeys, some are straight lines. Dänniken concluded that the straight lines were airfields and the other figures were messages that doltish humans were instructed to carve in the desert of Peru by extraterrestrial overseers.

Why? What's this about? We don't know how to draw big pictures without extraterrestrials telling us what to do? Some of those straight lines are six inches across. How big are the airplanes that land on those airfields? What are we to imagine, an interstellar spacecraft effortlessly crossing hundreds, thousands of light years, the cargo door opens, and out come little propeller-driven airplanes about the size my three and a half year old son plays with? This is silly. The common feature in all of von Dänniken's fantasizing is that he sells our ancestors short. He goes to Egypt; he sees pyramids. "Boy, those are big! How big are the constituent blocks?" "A hundred tons." "A hundred tons," says von Dänniken, in effect. "I couldn't lift a block that weighed half that." Therefore, human beings are unable to lift blocks of that mass; therefore, extraterrestrials did it. Q.E.D. But we have knowledge of how the pyramids were built. We know about quarries and logs as rollers and rafts up the Nile and inclined planes and tens of thousands of workers. We understand how humans could do it. Our ancestors were perfectly capable of building things big. The idea that extraterrestrial visitation is required every time a naive observer can't figure out how something could be done, is silly. The one possible positive aspect of von Dänniken is that, in frustration and perplexity, it drives an occasional reader into archeology.

Q: Happy Birthday. I'll be honest with you. A friend of mine, well, now an acquaintance of mine, sold me the idea that Steven Spielberg was going to be here, so that's why I came but look, I'm very happy I . . .

SAGAN: That's the first I heard of it. I'm awfully sorry that you're disappointed.

Q: Wait, I'm not done, I'm not done. I'm glad I came, because when you were doing that thing about pointing out where Earth is, and how little we are, compared to the universe, wouldn't it be easier for Man, women, humans, for humans to think that we're not the center of the universe if we have all this data? It's easier, why can't we believe, the heart of the challenge now is to think that we are in some way and somehow the center of the universe in some kind of way, not physically, not intellectually, but somehow, I don't want to make it like a purpose, but . . .

SAGAN: Yes you do.

Q: OK, it's a purpose. But it's now easier for us to believe, OK, it's easier, we can put our minds at rest, but somehow aren't we closing our mind by not saying that we're not in some way the center of the universe?

SAGAN: But imagine that you're an intelligent octopus.

Q: OK, I got it, but wait. What about organisms and the center of the universe. OK, I'll change it. Humans, because we can communicate and understand each other, I'm just saying ...

SAGAN: My mind is open. You come up with any evidence we're at the center of the universe, and I will gladly consider it.

Q: Hello. First of all, I would like to say I'm glad I didn't get to ask my first question about God, but my second question, which might be oversimplified, is, if every matter has an antimatter, and the Big Bang theory created the universe, which is supposedly matter, where is the antimatter in the Big Bang theory?

SAGAN: (Incidentally, you don't mean to suggest that the Big Bang *theory* created the universe, just the Big Bang.) There is quite sophisticated cosmological speculation and theory on this. Clearly, if there is an excess of one over the other, and the universe is well mixed, then, since matter and antimatter annihilate each other, whichever had a slight preponderance at the beginning would be all that's left. Einstein said, "Matter won." That's a possibility.

Q: You've described several major demotions that we have undergone. Do you think we're now as demoted as we can get?

SAGAN: No.

Q: What further humiliations can you see for us in the near future?

SAGAN: You see, the idea that our sense of self-worth comes not from anything that we've done, not from anything worthy, but by an accident of birth, is the crux of the humiliation, in my opinion. I would urge those of us worried about being demoted, those of us who wish for us to be important, to *do* something important. We should make easily understandable, achievable, and inspiring goals for the human species and then set out and accomplish them. That would give us the confidence that we sorely lack when our self-esteem depends on nothing we do. Let's make a planet in which nobody is starving. Let's make a planet in which men and women have equal access to power. Let's make a planet in which no ethnic group has privileges denied to other ethnic groups. Let's have a planet in which science and engineering is used for the benefit of everybody on the planet. Let's have a world in which we go to other worlds.

SCIENCE EDUCATION

12

Does Science Need to Be Popularized?

ANN DRUYAN
Federation of American Scientists

Does science need to be popularized? This is a tragic question, symptomatic of a dangerous illness that began to afflict our civilization soon after science was born. Of the first known scientists, the pre-Socratic philosophers of ancient Greece, Democritus, Empedocles, Hippocrates, to name a few, only fragmentary information survives. But a picture emerges of a regular bunch of guys. They got their hands dirty, making things. I am thinking of Empedocles and his use of kitchen utensils and plumbing scraps to invent the experimental method. They liked to get high (Democritus, especially). They didn't think of themselves as the elite members of a scientific priesthood, separate from the people. They were workers anxious to figure out how nature herself works.

Their sharpest tool was a revolutionary observation, best articulated by Hippocrates: People think that epilepsy is divine because they do not know what actually causes it. But someday we will understand what causes epilepsy and then we will no longer think it divine. And so it is with everything that we now call divine. This is what is known as the God of gaps.

If we consider divine anything we do not understand, God is a kind of shorthand for what we do not know, but an inadequate explanation for what is. We might well solve the mysteries of nature if we systematically set about trying to do so. This is really the wellspring and inspiration of the scientific method and the values that Carl has used his considerable gifts to communicate.

But this democratic notion of science belonging to everyone, this attitude of science being a normal human response to reality was to perish early on. The idea that you could investigate nature and find out how things were put together was distorted. Science became the property of people with wealth and with slaves, those who were too lazy to actually do the experiment, but preferred instead to lie back on their couches and make pronouncements about the structure of the heavens. People who thought that humanity should be divided into those lucky few worthy of doing science, the select who could know the elegance and beauty of nature, and everyone else who did the work that made their leisurely, contemplative lifestyles possible, but who were themselves, somehow ineligible for this information.

It may not be entirely just, but I enjoy blaming Plato for a lot of this. The idea that thinking should only be done in the Athenaeum or in some temple insulated from the rest of life, that is the reason we find ourselves asking today: Should science be popularized? It's because we think of science as something entirely separate from human activity. This is a grave error.

In that same place where science was born and in many of those same minds, the idea of democracy was also born, that dream that everyone, or almost everyone, should have a say in what happens and not be pushed around. This is to me a development of biological significance. We are primates, blissfully hierarchical, accustomed over millions of years to ceding most decision-making to one surly alpha male. Suddenly, around the fifth century B.C., there were those who no longer found this arrangement acceptable.

Science and democracy go back a long way together. They have much in common: Both depend on freedom of thought and freedom of expression. Arguments from authority have no weight. Just because someone with power says something is true does not make it true. The free exchange of ideas is the lifeblood of both systems.

We certainly have made tremendous scientific and technological progress in societies that were hardly democratic. We can have science without democracy. But I wonder if we can hope to have democracy without science. How can a citizen with little or no understanding of the methods, laws, and language of science hope to be an informed decision-maker in a society utterly dependent on science and high technology? If science belongs only to the few, how can the many hold them accountable? Thomas Jefferson worried about this, saying that those who hope to be both ignorant and free are hoping for what never was and what never will be.

Should science be popularized? You might as well ask: Should we have a democracy?

Science is the great baloney-detection kit that we desperately need. Why? Because our greatest strength as a species is also our greatest weakness. We are imaginative, but we are also terrible liars. We lie to ourselves every chance we get. We lie to keep power. We lie to keep other people from getting power. We lie to make ourselves feel special, to push away gnawing anxieties of mortality. We have a terrible fear that we're not central, that we're not being watched over by a loving parent who will protect us and who will help us in our fearfulness.

We cannot yet bear to accept our true circumstances: that we are tiny beings in a universe of incomprehensible vastness.

This tendency to self-deception means that we need a machine that will keep us honest, a voice whispering in our ear, saying, "Be careful now, you're very young, you're very ignorant, you're very new as a species. You thought this was true before. You were wrong. You might find out later that something else is true." We need that voice in our head, and that voice is science.

You don't want to impose your values, your fantasies, your expectations, on nature, because nature is so much more complicated than your best imaginings, your best dreams, could ever be. You want to know how it really is put together. That's the humility of science, really. I know science is often thought to be arrogant, and many of the practitioners of science are arrogant, but the greatest strength of science is that humility, that knowledge, that self-knowledge that we have to protect ourselves from this tendency to lie about what reality is really like.

There is a lot of resentment against science now, and we shouldn't idealize science, because in too many ways it has become a gun aimed at our heads, not for the well-being of humanity, not to make this planet a more habitable place, a more humane place. Science's worst offenses were largely committed in secrecy. There was no informed electorate to debate these decisions. Public mistrust of science is widespread, another urgent reason for an all-out campaign for scientific literacy.

I'd like to talk a little bit about the man I consider the single biggest contributor to this campaign ... I've had the privilege of accompanying Carl Sagan on nearly twenty of his sixty trips around the Sun. Countless times I've seen people come up to him on the street since *Cosmos* was first broadcast and never ceasing since that time, saying, "Dr. Sagan, I'm a scientist because of you." "Dr. Sagan, I was interested in science but I never thought I was capable of understanding it before I read that article in *PARADE*," or "I saw that episode of *Cosmos*," or "I read one of your books." They have thanked Carl for their vocations, avocations, and there is one refrain that I find particularly moving: "You made me realize that I was a part of the fabric of nature, part of the universe, a feeling that I yearned to have in church but never did." I often think of the porter at Union Station in Washington, D.C., who stopped Carl from trying to give him a tip by saying "Put away your money, Dr. Sagan. You gave me the universe. Now let me do something for you."

But Carl's efforts at education have not been confined to the general public. I remember with pride the courage Carl displayed before a sea of military fruit salad and hungry contractors with dollar signs where their eyes should have been, at the Department of Defense at the height of the Reagan–Bush Star Wars hysteria. Into the belly of the beast he went to debate General Abrahamson, the head of the administration's program. I think I was probably the only friendly face in the hall. Carl fearlessly, without anger or ad hominem, or straw men, without resorting to any of the rhetorical liberties taken by the opposition, debunked the Star Wars scam on the merits. It was a stunning performance, acknowledged by the standing ovation that followed.

I'm reminded of another miracle performed by Carl. In the 1980s he was asked to testify at an Arkansas trial over the teaching of evolution in the schools. Depositions were taken at the offices of a New York City law firm. On the interrogation team was a creationist expert from Little Rock. About a year later Carl received a long thank you letter from this person, saying that Carl had been so persuasive he could no longer promote creationism in good conscience. He had quit his full-time job with the creationists and was now teaching evolution in a small private school.

I remember the time General Alexei Leonov, the first human ever to walk in space, was introducing Carl's talk to the Society of Space Explorers in Washington. This is one of the most exclusive clubs in the world: you have to have flown in space to be a member. "Do you realize the debt that you owe to Carl Sagan?" Leonov asked. "He came to Moscow, to the Central Committee and he briefed them on Nuclear Winter. After he left, a dozen men on the General Staff looked around at each other, and they said, 'Well, it's all over, isn't it? The nuclear arms race doesn't make any sense any more, does it? We can't do this any more. The threat of massive retaliation isn't credible anymore. It jeopardizes too much of what is precious to us.'" Carl was tireless in his efforts to present the facts and implications of nuclear winter, speaking at the Defense Nuclear Agency, the National War College, all the service academies, the CIA, the staff of the Joint Chiefs, and a special closed session of Congress...

Carl undertook a concerted worldwide campaign to ensure that the governments and citizens of potential nuclear combatant nations as well as potential noncombatants knew what was at stake in thermonuclear war. He briefed the chiefs of government of Sweden, India, France, Greece, Canada, Tanzania, New Zealand, Mexico, and Argentina, among others, as well as Pope John Paul II. He spoke to parliamentary and general staff committees. He gave a series of talks and interviews in Japan. He wrote a petition cosigned by ninety-eight Nobel laureates calling for an urgent reversal to the nuclear arms race and distributed it to all the potentially and actually nuclear-armed nations.

A similar spirit can be seen in 1965 when, against the advice of scientific and political colleagues, he quit the Air Force Scientific Advisory Board in protest against U.S. action in Southeast Asia and in the late 1980s when he and I organized three of the largest public protests at the Nevada Nuclear Testing Facility against continued U.S. underground nuclear explosions in the face of a Soviet moratorium.

(They went ahead and blew up a weapon right in the middle of one of Carl's impassioned talks.) He was arrested twice.

The dream that you could have a society where everybody knew something about nature and the universe is one of our greatest innovations as a society, the best thing we ever came up with, and this institution, Cornell University, was founded by men who were consciously trying to realize that dream. They said every farmer should know something about the universe, every banker, every single person in our society. We are a democracy. That means that we are all decision-makers, not just the kings, not just the rich. So we should all need a good grounding in how the universe is put together. We should be practiced baloney-detectors if we want to have a functioning democracy, if we want to keep what little freedom we have. That started right here, in many ways, at Cornell, and other places in the United States, and I see so many signs lately that we're giving up on that dream, that we're giving up on our public schools, that we're giving up on our cities, that we're giving up on the people among us who have nothing, and saying, yes, maybe there should just be a class of people who do our dirty work for us. Why do they have to be educated? Quit educating them. What do they need to know science for? Well, if we give up on that original dream, I believe we'll become just another one of those dead empires, and there are plenty of them.

Science has many crimes to answer for, but it's just like our society. It's like the idea of democracy itself. It's imperfect, but what else do we have, to keep us from letting the worst parts of ourselves, those evolutionary tendencies that get out of hand every once in a while and shame us and cause so much destruction and kill so much of what is precious? What else do we have to keep us from doing those things, from committing those crimes? You're a great teacher, Carl, and the students who spoke yesterday attest mightily to the quality, the eloquence, the decency, the whole spectrum of what's best about human beings. I saw that yesterday, and you taught not just these students, but I would say a significant fraction of the people on this planet what science should really be about.

It is widely assumed that there must be some sort of inverse ratio between a distinguished career in science and personal development as a human being. This is another way in which Carl is exceptional. I want to pay tribute to Carl's parents, who, although no longer living, have played a tremendous role in who he is today. One of the speakers, Bruce Murray, said, "He can take a lot of punishment, that guy, and he keeps coming back." I know where he got that determination because I knew Rachel Gruber Sagan. She was a person of tremendous determination, and she adored Carl and believed in him. And his father, Sam Sagan, who by universal acclaim, was one of the most charming, one of the sweetest human beings I've ever known. I see Cari Sagan Greene, Carl's sister, here, and she's shaking her head "yes" because she knows how true this is. There's a striking similarity between Carl's parents and my own, Harry and Pearl Druyan, of whom I'm tremendously proud and who are here today. I think one of the reasons that Carl and I have such a deep affinity for each other is because we've been lucky to have such conscientious parents, who have taken their responsibility so seriously, to be strong links in the chain of generations.

I'd also like to say something about Carl's children, all of whom are at this symposium: Dorion Sagan, his first-born child, a talented writer, and father of ten-year-old Tonio Sagan, a gifted artist and a ray of sunshine. Dorion is the author of several books, many of them on science. His frequent collaborator is his mother, Lynn Margulis, who is also here today, and of whom we are also enormously proud. Lynn is a very brave person and a most distinguished scientist. And Jeremy Sagan, who has already made his mark in the world. As inventor of a leading computer program for composers, he has brought science and art together. Jeremy is now a Cornell student with straight As thus far. And Nick Sagan, who in his early twenties has already written two produced episodes of the *Star Trek* television series. To be twenty-two years old and gainfully employed as a dramatist is simply remarkable. Sasha Sagan, our twelve-year-old daughter, is both a fine writer and a promising actress, who got 109 on her last math test, and who has Carl's character truly, his integrity, his love of truth, his passion for finding out how things really are. Sam Sagan is only three years old, so we don't know exactly what he's going to be, but even at this tender age he has already begun to manifest a curiosity, an intelligence, a tenacity, an originality that we recognize as being Saganesque.

I'd like to conclude with a story of an incident that occurred soon after *Cosmos* was first broadcast. Carl and I were flying into customs at Kennedy Airport and you know how you are when you come into customs; even if you're innocent, you always have this feeling of, like, gee, you know, do I took nervous? Is there a little tightness around the mouth? Am I smiling too much? We were waiting on line, and suddenly an agent looked at Carl. "May I see your passport?" He takes the passport. "All right, Dr. Sagan, please come with us." Carl and I are thinking, what did we do? Did we declare the perfume? What is going on here? The customs agent calls someone else over, a supervisor, and we were thinking, well, this is going to be a long, ugly afternoon. They confer in secret for a moment, and then they step forward and they say, "Dr. Sagan, my partner and I don't think you've given Plato a fair deal. OK, you liked those pre-Socratics, fine, we like them too. Yes, but let's talk about some of the great things that Plato did, and let's think about him in the context of Aristotle."

It's just another sign of what can happen when you don't underestimate people, when you respect them and their intelligence, and when you want to share the things that mean the most to you, the ideas that you find most uplifting, most exciting, most stimulating.

Should science be popularized? In answering that question I would paraphrase the V'Ahavta prayer from Deuteronomy: Thou shalt teach it diligently unto thy children... Thou shalt speak of it when thou sittest in thy house and when thou walkest by the way... It shall be for frontlets between thine eyes... Not merely packaged as a collection of anecdotes and amazing facts, but as a way of thinking that you carry within yourself everywhere.

To one who has done this more effectively than any other... To the bravest and most decent person I have ever known and to the author of all my happiness, I say Happy Birthday.

— 13 —
Science and Pseudoscience

JAMES RANDI

Plantation, Florida

I am not encumbered with any academic degrees, and perhaps that is an advantage. I believe it is, to a certain extent. I am very much like a free spirit. I can call them the way I see them! For example, if a newspaper reporter were to appear on any given campus because a local professor had come out with something that sounds, and certainly seems in every respect to be, a little strange, but nonetheless does get some headlines and is featured on all the talk shows, so we know it must be important, the reporter might ask a colleague of this person, "Sir, can you tell me your candid opinion of the research recently announced in the press that has been done by Professor Brontosaurus?" and that colleague would probably reply in this general fashion: "Well, seeing that in actuality Professor Brontosaurus, who as you know has been published in many leading scientific journals and whose work of course I support, as do all of my colleagues – make a note, please – is an acute observer of nature and a great scholar, and a man who has tenure at our university – are you making notes? Good. Though one might presume that he has a relatively small data base upon which to base his present stated conclusions – not that that data base is insufficient per se, but it could be expanded perhaps in order to give a better statistical picture, you understand," and it goes on and on like this and the man never gets to say anything. The same reporter comes to me and says, "Mr. Randi, what do you think of the announcement of Professor Brontosaurus?" and I say, "Frankly, I think the man is not rowing with both oars in the water."

I am privileged to be able to say that, you see. I am in a most peculiar profession. I have traveled all over the world, and I have appeared before academic groups and groups of lay persons as well, and mixtures thereof, all over the world, and I tell them things they should already know. It's a rather thankless profession, to a certain extent. I will appear before audiences and tell them things like, oh, for example, Richard Nixon did know what happened with the Watergate tapes. Carl Sagan never actually said, in so many words, "Play it again, Sam," or whatever it was. And that Chicken Little was wrong; the sky is not falling, you can be reassured; and Elvis is really dead. Going about the world, telling people things like that, doesn't make you terribly popular. People are still fussing about cold fusion, which in my opinion and the opinion of many of my colleagues probably just does not work, but it does work in one respect. It gets a lot of funding, at least from Toyota, who just gave them $7 million to pursue cold fusion studies. Wonderful! I must also announce a distressing bit of news that I am currently arguing with my very good friend, Arthur C. Clarke, in Sri Lanka. I'm glad that he is at a considerable distance from me. We might be in a fistfight, because he is quite supportive of cold fusion. He has spoken to the founders of this wonderful notion and is pretty well convinced by them, so I may have to go over there and clast that icon too!

I speak to a great number of lay audiences and academic audiences, and we have to get some terminology straight. Pseudoscience and crackpot science are differentiated in certain ways. Examples of pseudoscience in my estimation are things like homeopathy, which is diluting a medicine down to the point where you're beyond Avogadro's Limit, and there's none of the original medicine left, but the vibrations are still there. Some parapsychology, in fact, I think most of parapsychology, is also pseudoscience because of the way it is approached, but parapsychology is a legitimate science, no question of that, and it must be pursued. It is in an unfortunate position. It's been around for something like a hundred and twenty years, not necessarily under the name, parapsychology, but scientific research directed in that way has been around for that amount of time. When I speak to parapsychologists, they usually say, "Well, I still have a feeling there is something there," in spite of the fact that they have not had one positive experiment yet, in more than a century, that has been replicated. Strange! It is very much like, in my estimation, being a doctor for 120 years, and every one of your patients has died. After the first thirty years, wouldn't you get an idea that maybe you should seek a different line of work? Psychic archeology is a thing that is coming back for attention, and I am not going to get into that. Brain wave research in Russia is also pseudoscience, though there is obviously excellent research going on in that country and has been for quite some time now. They really have accomplished a great deal in certain directions. In parapsychology, however, they have accomplished no more than anyone else in the rest of the world has accomplished, which is zero.

But pseudoscience is equipped to look like science. That is what makes it different from crackpot science. They have lab coats – that is very important – preferably not too clean, with a few acid holes burned in them. They publish long papers, 40% of the paper will be footnotes. That makes it legitimate, and there are always more references than text; that's important. There are lots of signers, by people who may not know anything about the subject, but they have tenure. There is lots of instrumentation. In fact, it is getting more and more involved. The instrumentation that we saw in Russia was exceedingly large in quantity, but applied in the strangest ways. They wired me up, head to foot, with all kinds of little sensors, and said that if I looked at the screen on this Apple computer, that I would probably be able to make the dot move, and boy, I could. I could make it move by holding my breath, or by squeezing the leads, surreptitiously, of course, behind my back, in the tricky way that we conjurors have, and by leaning forward in my chair. Several things caused that dot to move. I could have written my name on the screen with the dot, and they were convinced that I had marvelous psychic powers. The whole thing was turned up so strongly that when a car went by in the alleyway, it went right off the scale. That is pseudoscience.

But then, crackpot science. That is a much richer field. Neither pseudoscience nor crackpot science has any legitimate findings, but occasionally from pseudoscience there develops a real science, but for crackpot science there is not much hope. Facilitated communication is the latest rage in psychology, along with these recalled memories from childhood of tossing babies around on bayonets and drinking their blood and various things like this – absolute, total nonsense, of course, but very, very damaging.

Facilitated communication, for those of you who may not be aware of this latest stunning development in science, is taking the hand of an autistic child and guiding it around a keyboard to type out answers. Unfortunately, some tests that I did at the University of Wisconsin at Madison, sort of indicated that there was nothing to it, at least on that occasion at that time under those circumstances with that sample. I sat there with some autistic children, watching these facilitators holding their hands and typing out answers, very well spelled and very well thought-out answers, and unfortunately the children spelled only as well as the communicators spelled. I thought there might be a connection there someplace, and it turned out that when the communicator did not know what the answer was supposed to be, the child also did not know it.

I had actually been called in there in the first place, believe it or not, to find out whether or not the autistic children were telepathic. I jumped at the chance, of course, and the problem was that there was an autistic child on one side of the room who had not met the autistic child on the other side of the room; nonetheless, if the facilitator went to this child and said, "Think of the word basket, basket," and show it to him in writing, "You see? Basket!" then would cross over here and say, "What is that child thinking of? Let me have your hand," and the child would tap out, "Basket." Isn't that astonishing? I have no explanation for it whatsoever!

Faith healing is another example of crackpot science, where they apparently come up with lots of evidence, anecdotal, and when actually tested it does not work. Perpetual motion, free energy. You have to be very careful with perpetual motion, and that is because they don't want to call it perpetual motion. They know that doesn't work, so they call it free energy. You see, you set the machine going, and you connect up to a machine that charges the batteries that run the machine. You follow that? It's called circular reasoning. A man named Joe Newman raised $17 million, partly through the Mississippi Board of Transport and Energy and he got the signatures of thirty Ph.D.s to verify that his machine indeed does put out more energy than is put into it. He is currently suing the Patent Office in Washington and several other people for not recognizing the worth of his wonderful invention.

Creation science, of course, is a perfect example of crackpot science. Now, crackpot science depends upon a few things to be successful. First of all, appeal to authority: Holy Writ, divinely inspired documents of various kinds, or for example, Joe Newman appeals to Maxwell. In his early days, when Maxwell was sixteen, he wrote some sort of a short essay on magnetic vortexes and later withdrew it when he thought more about it, and damned his own writing very reasonably and very correctly, but Joe Newman says he was forced to do that, of course, by the oil companies, so he depends upon the writings of Maxwell when he was only sixteen.

In parapsychology we find special pleading and special exemptions all the time. One of the supreme pieces of reasoning that was brought up some time ago in the ESP field (i.e., in the parapsychology field) was originated in Bath, England, by two sociologists there who were parapsychologists in their spare time. They came up with this wonderful piece of reasoning. They said that since psychic forces work backward and forward in time, and don't have any falloff, the law of inverse squares doesn't apply to ESP. It works equally well over any distance, you have perfect transmission, since it works forward and backward in time, and that was the reason why there were no positive research papers published on parapsychology. They reasoned thus: If they did some research that proved that these powers exist, they could then write a paper, the paper could be published in a scientific journal, and then someone like James Randi would read it. He would send out negative vibrations that would go back in time to the time of the experiment and the experiment would fail. Now, whether that would change the results in the published paper or whether there would just be blank places in the journal, I am not sure, but it is something to think about, isn't it? Not for very long. I wouldn't waste too much energy on it.

The crackpot scientists also appeal to secret substances and devices that they cannot reveal to you, locked boxes, things with sealing wax on them, and such, and the overpowering theory that they have, or requirement, I should say, is that their theory must be believed first and then witnessed in action. If you believe it first, that is supposed to equip you in some way to be able to observe the results better.

Now, before I go too much further, I should relate the following. I have received a great number of questions about this gentleman, Uri Geller, who came from Israel some years ago and went to Stanford Research Institute in California and caused a great sensation when apparently he could bend spoons with the power of his mind. Now, there's a profession for you! He is currently suing me for $30 million, which is not a realistic figure to sue me for. Well, he's losing the lawsuit, I'm happy to announce, and is preparing to settle it all, and that is a happy circumstance for me. I must say that since he claims that he is bending spoons by divine powers, I have discovered many ways, as have all the conjurors, magicians, fakes, if you will, to do it without divine powers and just as effectively in every way. In fact, when we see him doing it on videotape, it's impossible to distinguish between the divine method and the trick method. So, if he's doing it by divine means, I can only tell him this: "Mr. Geller, you're doing it the hard way."

Now, why do we have crackpot and pseudoscience? There are several reasons for it. First of all, publishers just adore books that support these things. *The Secret Life of Plants* made millions for a man named Cleve Backster some years ago. That was where you hook up a plant (a *Dieffenbachia* is rather nice to use) with a polygraph, put little electrodes on it, and then you stand there and you threaten to burn it, and the polygraph goes off the scale. Isn't that wonderful? I didn't think that plants had central nervous systems, but apparently I'm wrong on that assumption, and all the botanists are wrong too, but that got a lot of publicity. As a matter of fact, it didn't fall apart really in the public estimation until he did his final experiments. And again, don't blame me for this, this is his experiment, not mine: He found out that if you take two containers of yogurt (please don't laugh, this is Science) and connect them with a wire, and then you dip a lit cigarette, which is a waste of good yogurt, into one of the containers of yogurt, the other yogurt will react to the polygraph, but only if both containers of yogurt came from the same culture. That's very important. And that was his conclusion, believe it or not. At that point, he lost some following in the scientific community!

Publishers are fond, as I say, of books like that, and on the lecture circuit, Jeane Dixon obtains very large fees. She gets $20,000 a lecture, and her lecture halls are filled to overflowing. They turn people away, all the time. Now, let's look into her record very briefly. First of all, in a five-year period she made 364 predictions that were dutifully recorded. Those are the ones that were published, not ones that she mumbled at cocktail parties. Three hundred and sixty-four of them. Guess how many were right? Of 364 over a five-year period, four were correct, and one of those was great medical breakthroughs this year! Wow! How ever did she know? Now, there's genius for you. Four of 364 is slightly better than 1%. I don't think it's enough to make a difference at the racetrack. But Jeane Dixon is still the greatest prophet of all times. She says she predicted the assassination of Kennedy, right? Wrong. She said, first of all, "He'll never run for the position of President of the United States." He decided to run. She said, "Well, he won't win," and then he won, and she said, "He will probably die, either in office or shortly after leaving office," because she was going along with the twenty-year curse that afflicted all of these presidents!

In television, we just had announcements in the media press that we have two new proparanormal series coming up on the major networks this year, and everyone will eat them up, of course. Libraries? Oh, libraries are very selective about what they will put on their shelves. In Vancouver, Canada, recently, I counted forty-one pro-UFO titles, not one against. Not one. Nothing by Phil Klass, nothing by James Oberg, nothing by people who have seriously criticized UFO so-called research, but forty-one saying that it was absolutely true. In Washington, D.C., none of my books is in the library system. None of them whatsoever. And when I asked about it, it was said that "Mr. Randi has a bad attitude." Yes, I do. There's lots of Edgar Cayce there. He is the man who predicted that New York would sink below the waves about twenty years ago. It didn't, so that book is no longer in circulation. We take it out of circulation; it doesn't quite fit.

On TV, there is a program called *Obscure and Erroneous Mysteries* – I am sorry, that was someone else's designation of *Unsolved Mysteries*. *Unsolved Mysteries* calls me regularly. They only call me when it's a new researcher who doesn't know who I am. They call me up and say, "We'd like some information on so and so," and I say, "Oh yes, that's been solved. It's such and such." And there's a long silence on the phone, and then, "I don't think, Mr. Randi, we can use that opinion. I'm sorry." Why not? It's *Unsolved Mysteries*, remember? They only want them if they're unsolved. So they don't call me much any more. A couple of years ago, *20/20* called me. They were doing a piece on psychic surgery. I spent over four hours on the telephone with them, sent them several hours of videotape, and endless amounts of material. I watched the program a few weeks after that; they didn't use one word of what I sent them, but they used all the propsychic surgery stuff. They wanted it to work, because that's what people want to hear. Whether it's true or not doesn't make much difference to them.

I used to work for *Omni* magazine, used to write columns for them and special pieces. Then, the word went out from the editor-in-chief saying that proparanormal stuff was much better than antiparanormal stuff, and no skeptical attitudes from now on. The next issue came out with Telepathic Frogs' Legs on the cover. I figured that was too much and I resigned.

Now, some of the things that I deal with are rather disappointing. Sometimes I have to destroy peoples' belief in something, or at least cast some doubt on them. They are like rubber ducks, though; they always jump back. There is nothing you can do to discourage the true believer. For the moment, they may say, yes, well, maybe I don't have this aura or whatever, but you give them a couple of days and they will bounce right back. My friend, Dick Smith, who is a millionaire very well known in Australia, sponsored some tests of water dowsing some years ago. While we were there we designed a series of tests, as I have done in many countries around the world, to test the forked stick or the pendulum or the coat hanger wires or whatever. Some people do it with their hands. And we did it in Kassel, Germany, two years ago, a very definitive set of tests, and, of course, it proved that the law of averages works quite well, but dowsing doesn't. With the tests that we did with Dick Smith in Australia, we had eleven dowsers and we stood them all up in the office under the video camera, and Dick asked them, "Now that all of you have failed when you said that all of you would get 100% results, do you have any changed opinions? I'd like to see, by a showing of hands, how many people still believe that they have the ability to dowse?" Ten of the eleven people held up their hands, and one didn't. We asked him why not, and he said, "Well, I'd have to think more about this. I have my doubts now," and Dick was ecstatic. He said, "One of eleven. That's the best score we have ever had in trying to convert people. In fact, I have never seen anyone converted!" and I said, "Dick, give them a week or so." Well, it wasn't a week; it was the next day. Dick called me up and he said "Yes, he decided that it was because Jupiter was in Sagittarius and that is always a bad sign for him." Incidentally, they did replicate those tests again in Perth, in another part of Australia, with the people who got the highest scores, which were still not high enough to qualify and they failed even more miserably.

Now, I sometimes get the question, why do you do this, Mr. Randi? Well, I do it because I'm trying to inform. I think that that is what the job of an instructor should be, trying to inform. I feel very much like the man who sees somebody struck by a car, and because he is not a doctor, has no medical training, wonders what he should do. Well, I think the reasonable thing is to drag that person out of the path of the traffic and call for medical help, and that is essentially what I am doing. I am saying, perhaps you are being swindled, or deceived, or perhaps you have got a wrong idea that I might give you a different slant on. Now, if that person wants to crawl back and play in the traffic and get run over a second time, that is okay. Maybe I will drag him back a second time, but not a third time, believe me. After all, that person has decided to play in the traffic and be damaged. All I do is offer my help.

But why is magic so very popular? Why is magical thinking so very popular? Well, for a few reasons. First of all, religion, to me, is one form of that magical thinking. It has easy answers, and they're positive answers. They don't have ifs and buts or percentages or statistical probabilities. They're absolutely positive, 100%, this is sure, and will always be, always has been, amen. There is no critical thinking involved. It is already written out in the book. You don't have to do any thinking. You don't have to judge it or evaluate it. It's there. They're saying, "I want some rules I can live by, and rules that I know are absolutely determined. I don't want to have to think about it." One hundred percent certainty is what people are looking for, and our world isn't made up that way. It just doesn't work that way; it never has, and it never will. All we do is try to get the best percentage we can out of it.

I had a woman in an audience one time. The dear soul was sitting in the second row and when I came to the end of my lecture and I said, "At this point we will have questions and hopefully, answers." She held up her hand and I said, "Yes, ma'am." She stood up and she recited something that was said to have happened to a relative of hers in Scotland many years ago, and I said, "Well, I wasn't there, you weren't there, it happened in 1902, it is rather far back in time, but I can give you an opinion as to what might have happened under these circumstances," and I proceeded to do so. She sort of nodded, looked a little puzzled, and sat down, but immediately stood up again. I said "Yes, you have another question?" She said, "I figured out what your problem is, Mr. Randi." I said, "Well, don't tell everybody! I have a number of them that I wouldn't like to have aired!" She said, "Oh no, your basic problem, I've solved what it is." I said, "Well, then, do share that with us." She said, "You're overobsessed with reality." Now the audience laughed, and she looked around and smiled, as though to say, "We got him, didn't we?" I paused for a minute and I said, "I'm thinking, I'm thinking! Yes, I think, ma'am, you're probably absolutely correct." And she said, "I thought so," and she smiled, and the audience laughed again. I'm sure that that woman is still telling the story that she put James Randi in his place.

The next speaker, or questioner, I should say, stood up in the back of the audience, and I said, "Yes?" and this man said, "Ah, before you leave this evening, I think the most important question is, can you give us a good lottery number?" Sometimes you just don't get through to your audiences, you know?

In closing, I must say there is hope. There is a lot of hope, of course. I have given you a lot of negative aspects of what happens in the world of pseudoscience and crackpot science and misinformation. The media have, by and large, deserted us. They have gone for sensationalism because it sells. It gives them ratings and it sells the books, but there is a lot of hope. Such things as the new book called *How to Think about Weird Things*. It is meant for freshman and sophomore audiences, and it is really excellent.

Then we have the Committee for Scientific Investigation of Claims of the Paranormal, and that is one of many skeptical organizations all over the world today who are doing a very active job in trying to act as information sources to the press, to the media, and to the public, and they are doing a wonderful job. Their journals are distributed all over the world, in very large circulation. Many libraries carry them – not in Washington, D.C., though, you can bet on that. So, there is hope in that respect.

The *Nova* program itself on PBS – I practically worship that program; I cannot tell you what a feeling it is to see the *Nova* logo come up on the screen on the day that your program is being broadcast and know that they are going to deal with you. What a thrill! Nothing has surpassed that in my whole life. It really made my day; it made my life.

And there is something else going for us. I am sure you know what it is. It's Carl Sagan.

14
Science Education in a Democracy

PHILIP MORRISON
Massachusetts Institute of Technology

Experts and Freeholders

Close to fifty years ago a gifted anthropologist came to Cornell: Allan Holmberg. He became a friend and rather a hero of mine. Not many years later he died in the midst of his work. He wrote one beautiful ethnographic book, among many papers; his 1950 book was named *Nomads of the Long Bow*. It told of an extraordinarily isolated and strangely placed people, hunter-gatherers in the wet Bolivian forest at the extreme headwaters of the Amazon. It was clear from the tales they told and from what their neighbors knew that they had been pushed there, a few thousand people in a hundred bands over a couple of centuries wandering through the rain forest into worse and worse circumstances: more swamp and less land, much less protein, much harder living, until they were foraging at the hungry margin.

Of course, they were by then highly skilled; they had managed a ciphered speech made of bird calls, so that they could communicate surreptitiously in the presence of adversaries or of timid prey. From their canoes on the rivers the early Jesuit explorers glimpsed these people fleetingly and heard the bird calls, but never a word, or any familiar cry or call, and conjectured that these folk had no language. That claim inspired Holmberg as a graduate student at Yale to go out to seek these unique people; what, he asked, were their dreams?

It was of course a bizarre notion that these were humans without language, who spoke only in bird calls. That was only when somebody else – who didn't know their real speech – was listening. By this century they had made some outside contacts; they worked as itinerants in the forest-edge plantations whenever they had to. Holmberg made up his mind to visit them, and he did. On two occasions he spent a year more or less among the people in the swamps, first to do the field investigation that formed his thesis, and a year or two later as a consequence of World War II.

The place was remote; when Alan first stayed there he heard of the war in Europe only six months late. The only strangers to enter their territory in some numbers were the rubber tappers attracted by the bonus pay for natural rubber, then under intense wartime demand. They sought out every far corner of the rain forest. Unfortunately, these hard-boiled frontiersmen, trying to make a risky wilderness living, were not tolerant of small silent people with poison darts and tended to shoot them at sight.

Of course this precipitated open war between the two foraging groups, an abhorrent war that even came to impede the rubber collection. Young Holmberg accepted the charge to go back into the forest, make contact with his friends the Siriono, whose language he spoke by that time, and try to broker a truce. They would no longer be shot at, if in turn they would blow no darts at the rubber hunters. The truce worked out rather well.

The Siriono had an amazing tradition, which I must summarize in just a few sentences. They told the story, quite believable, that human beings had once known how to make fire, but they had long forgotten! Well, of course, they were the very humans who had forgotten, or at least put the old craft aside as impractical, because they lived in circumstances where most plant material was soaking wet all the time. Fire was all but impossible to raise by fire drill or fire plough; they could most easily have fire simply by keeping it going forever.

Their need had nothing whatever to do with hunting of big game, or with bonfires to frighten off predators. Fire helped feed them by extending the range of what could be eaten, especially proteins. Indeed, we all share that use; raw meat is not a staple for most. These people in a protein-short and watery world prized anything proteinaceous they could find, even if it was not very palatable or weighed only a few grams. Insects and their egg cases, cobwebs, even shed skins could be made digestible, even palatable, by denaturing all the tough and unusual proteins.

Fire was therefore of major importance to them; no fire, not much good food. Of course, they would not starve should they lose their own perpetual fire, because in the course of their wanderings soon enough they would find another band to share fire with them. There was only one profession among the Siriono: specialized firekeeper, who held tiny glowing embers in a long firebrand. Usually they were respectable, able-bodied women, who could be relied upon to do well what everyone needed. As long as your firekeeper did not fall into deep water, your group could count on usable fire. Theirs was a sense of true responsibility, and to fulfill it they held a special technique, no very elaborate one but important as a high duty. The shaman has a complex and long history among many foraging people, as psychiatrist, healer, wildlife manager, priest; an early division of labor, all specialization in the hands of a certain few men or women. But the Siriono had no shamans, no experts at all, save for their fire-tenders.

Thomas Jefferson's own model of his nation – ours now – was striking. He was utterly won by the independence of the sturdy, self-reliant, farm family, then most of the Americans. The freeholder and his wife dwelt on their own private land with children and family and livestock, a long-matured grainfield rotation, a woodlot for fuel, and the forest nearby for construction timber, beholden to no one. They traded a little of the farm yield for salt, pottery, ironwork, fibers, and the services of the grist mill, maybe a few books and newspapers as well, all the necessities the homestead did not make, and they accepted tithing by the church and taxes by the state. This freeholder backbone is the image of the Jeffersonian account of the nation. Somehow America – never mind the cities and the factories that were so clearly growing – would be the land of the independent. Jefferson was wrong about the growth of industry, a process he expected to remain mainly in Europe; and slavery too set his vision awry.

But it was schooling in symbols, words, and text that was the thirst of the freeholder. That's the one thing the family could not make for itself. It could not improvise the Bible or the Constitution on the farmstead. Life and death, growth and change, the Moon and the stars, the streams in flood and ebb and icebound, the seasons, the green and the wintery lands they knew splendidly from rich experience. The lore of nature was fixed in the memory of the old and by rich daily experience for the young; the young knew well what was commonplace, and the old knew how to cope with the infrequent emergencies across a hundred years of familial and neighborly memory. Theirs was a remarkable way of life, with few specialists.

The Three Rs

They lacked symbols. Therefore they sought schooling in reading, writing, and a little arithmetic; those were what schools were meant to teach. Reading and writing were in the service of the state and of religion, two great systems of belief not forced upon the people, yet accepted by most. They could be realized only as the schools taught the written and printed word. Some buying and selling and the counts of votes and taxes meant arithmetic for all. Those were the tasks of Abe Lincoln's school, the school of the young republic, and the task for which schools were founded, at least across the farmlands where most Americans lived and worked.

In the small world of the Siriono everyone knew the specialist, the firekeeper, personally. The indispensable expert know-how was at hand and not at all that difficult. Among the million families of

Jefferson's America, there was also not much special knowledge the freeholders lacked, save for the work of clergy and lawyers, and maybe the local smith. Reading brought the hardest part of it in essence to everyone, though few could visit the capital or know any but the local legislator. The specialist's knowledge was deemed important and opened at entry level to almost all (again, slavery denied the democratic dream). A few would persist to become smiths or millers, country lawyers, or clergy, only a few.

But today we live in a world of high science and its applications high and low, whose many adepts can and do change our lives on a short time scale. A democracy can hope for wide and informed consent only if that special experience and knowledge is also opened widely at entry level. It is not the changing specifics that are the most urgent to share; it is the means by which adepts arrive at their new and credible specifics that is the urgent and enduring lesson.

Symbol-Rich; Experience-Poor

We live in a very different world now, a world where at megaHertz rates images flash by the hours out of a hundred million television screens, where printed paper drifts over the land like autumn leaves, and then the cables and dishes bring still more symbols into very many households. We have a tide of cheap symbols. Symbols would not seem to be so much in demand as they were in Jefferson's days, but in fact, the schools are still bound to symbolic instruction. That is their main task as they see it.

The conservative view of education insists on that central task in and out of the schools, because symbols are so useful and so cheap. That is the main reason for the present-day pervasiveness. If it costs a dollar or two to send a message a thousand light years, as Paul Horowitz told us, imagine how little it costs each viewer to see those wonderful images that Carl can exhibit to twenty million sets at once. It costs very little and it is strangely new. I believe that is an essential element behind widespread recognition that the schools are not succeeding to meet present demands. They spread not experience, but symbols, now in surplus.

Science, quite plainly, is now salient in the society, not because it controls thinking nor because it controls the way people behave, but because current technology, increasingly founded upon science, implies social and economic change.

It is rapid change that is the principal problem of this, our fast-changing time. What you learned ten or twenty years ago is not so serviceable today. The foundations of everyday life, social, economic, and technical, change rapidly and you're strongly induced to take part in these changes. Of course those changes imply a string of vital decisions.

Societal decisions, arrived at in a democracy by the consent of the governed, will not be reached primarily by teaching the shifting substance of every conflict. That substance, what kind of power plant to have, or what drug to dispense, is complicated, specific, detailed, and not even well understood at the time public decision is sought. Something else has to be done; we must strike deeper. Most people might agree with this. What we need is a deeper examination of the matter, of its foundations, of the nature of the claims, and done widely in the schools.

Social, technical, and economic change can be responded to by what I would call by a somewhat pretentious term, applied epistemology. I think this is pretty much what people have been talking about in the last days here. Put in much simpler words, how do we know what we know? It is the evidence and the probable inference from the evidence that are the subject matter of science.

Unless you know *something*, you can't exemplify evidence, you can't say how we know what we know. That is not a merely methodological study. It must be done through the substance itself, through examples. There is a need of some specimen on the table, a frog or a machine or a table of data or even an idea, but something to debate and to examine for itself. That's what I think we have to speak for: the concrete *sine qua non*.

That is also an experience. It is participatory, consistent not only with our democratic views, but consistent with an increasing body of understanding about human beings that we learn not only, perhaps not even primarily, through careful formulations clearly repeated. No, we require a richer experience: some of it touches on affect. We should like the people, like the words, make humor about it, enjoy the circumstances, wonderful sights, or sounds or motions or touches. What some people call the flow of the events must itself bring a reward, not simply that excellent moral nugget at the end where you've learned something.

We need more. That's why I don't think you can approach it well in any other way. You have to have your hands on, eyes on, mind on, you become friends with some stuff. It might be quite abstract. I don't say that it could not be prime numbers; I don't say that it could not be the day's rubbish. Those are both adequate vehicles for an epistemological approach: how we know what we know. But some vehicle we must have.

The Stuff Itself

Science has one wonderful property, imperfectly. (Its imperfections are its most interesting portions.) It casts a very wide net. You can go from almost any node of the net to any other node, even if by a circuitous path; even if you fall through at one place or another, those gaps will be where novelty and interest lie. I am not speaking for one well-defined curriculum; content can shift with the circumstances of the day, with what people want, with places and times and social conditions, changing as opportunity changes and interests change.

The stars, even the daytime sky, are wonderful to study. An eclipse is a fabulous phenomenon, though more people view the eclipses in America on the television screen than they do in the sky above. That's a victory by the physicians in the interests of public health, they tell me: I hope they are right. Some numbers of persons receive permanent retinal damage, some obscuration of the visual field, in every eclipse, because they look too long at the direct sun. If you have millions of people, you can affect this by getting them to stay inside, so you feel it worthwhile. (I concede it is arguably worthwhile; I am not persuaded that we are not paying higher if subtler costs by losing the sight of dramatic reality.)

Let all sorts of things enter, the school experience, eclipses to radio telescopes, chrysalis to DNA chromatography, depending on the age and the interest. One urgent matter comes with them, something the schools place very high, though I wish to modify their use of it. Yes, certification of the learning of effective behavior is essential, but that certification is not best performed by choosing little squares in a test, nor even by some cleverer test. It should be more nearly internal to the participant.

That's one feature of all sciences. Many times you can confirm your inference by noting confirmatory or disconfirmatory experience. That can be unforgettable for young people. Even among mature scientists, I think that sense of internal certification is urgent: the circuit works, the bud soon opened, the color appeared, the squirrel jumped ... A million events could be listed, all much sharper and more delightful than marking with an x the box that describes true typical behavior. Unless we do that better and better, we'll not translate the subsoil of science on large scale.

Two Goals for Science – and Also for Education

In the first sentences of one of Niels Bohr's thoughtful essays he points out that science has not one goal, but two. It seeks *both* to order the world and to widen human experience.

Most people have forgotten that. They say, and it is true, that science has the goal of ordering the world. But so has the philosopher, so has the poet, and the artists; the difference is primarily that the true order of the world is not a given, a passive matter of putting the things that are into the right bins. That was pretty good long ago when the bins were few and the tumult of a working world was great. Now the bins are very numerous and the tumult of the world has been enormously increased by the very content of previously ordered bins. What are radio and television and their data streams but consequences of Maxwellian, Hertzian, Helmholtzian physics of 120 years ago?

Aristotle's world had fewer experiences to order. You cannot produce the experiences we now swim in – they have grown prodigiously important and powerful – you can't produce that by ordering even the old world of 1810. No matter what philosopher looks at it, he will not grasp the issues. Jules Verne, who lived at the edge of it, saw there was going to be something like universal telegraphy that would bring text into the home, but even he could not imagine the worldwide video link we now see.

Order is in fact active. It must be able to fit novelty. That novelty tests the agreed order. It tests previous order. It determines the choice of theories, whether they stand or fall by whether the new thing not known when the bins were labeled can be accommodated. That surely is the main story of the feedback from theory to experiment at the level of high science, and I think it is to be found in all levels of experience.

People can understand that air too is a fuel, a fuel absolutely indispensable to the plants. Of ten pounds weight of plant, less than one ounce comes from the soil. For a hundred pounds of gasoline burned in the engine, another 150 pounds or so come out of the air. The consequences of what that does to the general ocean of air are increasingly important public matters, still widely misunderstood. They are even ignored, exactly in the way that science in its infancy had to ignore them, until during the seventeenth and eighteenth centuries it became quite recognizable that air had weight, air had substance, air had chemistry, air was an important actor. Indeed, the explanation of the work of our lungs was as recent as that. The Greeks had very strange and untenable theories about what the lungs might be doing. Air: you feel something, but it's not a substance, is it? That's the sort of learning that by early participatory experience would become a foundation for public policy.

Take a completely different domain. Everywhere from Halifax to Patagonia we Americans can say we live in the New World. Now, we don't live in the New World, frankly, because of what Columbus did, or Leif the Lucky, for that matter. That recent contact is quite another tale. It's an important part of European history, yes, but it's not the reason we can call ourselves the New World.

We are the New World for a clear, qualitative reason, one that doesn't need a technical argument to date when those people came, by one or another route, from Asia into the New World. The names New World and Old World are not themselves of importance, but the facts I believe are. A hundred and more of old sites have been excavated in the New World, where we find old tools, the old bones of humans and animals, walls ornamented, marked by occasional pictographs and engravings. Not one credible find has shown the handiwork of the ancient predecessors of all the humans now alive. Not one site known in all the Americas offers a clear sample of the work of the truly old people of the other species before *Homo sapiens*. In ancient times, when Africa, Asia, and then Europe were homes to our earliest protohuman forebears, the Americas had none.

When the ancestors of all the Native Americans came to the Americas it was a world that had never heard language, known tools, or seen any version of human society. It was their New World, the prehistoric New World. Australia has almost the same history. The oldest dates for human occupation of Australia are now somewhat earlier than the oldest credible dates for America, about forty or fifty thousand years. The first Australians were also *Homo sapiens*, as all humans are today in all continents. The reason is pretty clear: The falling sea levels of glacial times came in the times of our sapient ancestors. They could manage travel by canoe or alongshore. That was enough to get us where we are today, on all continents, where the earlier forms of hominids could and did know only the wider areas of the land-rich hemisphere.

I don't know how to make archeology very participatory for the schools, but I think it can be done. In fortunate places there's something to work on, and a very vigorous activity for reconstruction is no bad thing, if you get in there and work at it. You've got a trash pile, and you get to sort it, and the thoughtful producers have put an arrowhead or two in to make it more interesting. If that is done in an honest and appropriate way, I think it is worthwhile. You can also make your own stone tools by deft skill, still more interesting.

On Perception

Let me offer a newer topic still for schooling. It goes with my statement that most of science is the question of how we know what we know. The beginning of that is knowing how we know with our senses. Of course, that's very difficult, now an elaborate frontier activity for science, approaching remarkable growth because of modern developments. We need more work on human perception, the rudiments of it. Extending that to the senior high school domain, we reach the synthetic domain of digital image processing, which is going to turn the visual image into fiction as surely as the fiction you write on the white page or screen in symbols. Without that extraordinary visual reminder that those English words are not the adventure, not the ship, not the wind, not the child, but only words about them, even the most experienced reader would lose that distancing, at least a little. That's why she's free to stop, consult another page, even open another book.

None of those things happen in the continuous flow of moving images, and none of them at all happen when you're confronted with images that have semblance of continuity. If we can seamlessly create those myriad finite spots digitally, pixel by pixel, to suit our needs, we are facing a new world.

At the moment, we are in transition. It's still true that if you see a photograph of Gettysburg, and you have some belief in the authenticity of the path, always necessary (forgery is nothing new to our day), you can recognize things in the photograph the cameraman did not expect you to see: the way the tents were made, or what is happening in the background. But once you see the Lucasfilm video, and Marilyn Monroe and Abraham Lincoln are cavorting down the walk, you know that there's something wrong with that image.

That's only the crudest beginning, perhaps five years old. In ten or fifteen years' time, I'm afraid the countrywide video will carry half-synthesized and half-real images. When it is not deliberately misused, but simply enters into the currency of common view, the only way to protect yourself is what skillful teachers have used for a very long time to do the same thing about books. A sophisticated audience knows that books are written by its colleagues, and often wrong, and can be improved, and will be improved in the next edition. But that is not the case for the school child. The good school child very often thinks the book is telling the truth. It's in the book; it can't be wrong, or it surely is not likely to be wrong.

The best remedy against that is to have those kids make a book of their own and then discuss it six months later to see what's true and what's false. It brings a clear sense of the human quality of publishing books, the decisions, the choices, the sense of integrity that is required. Now it is the same story with moving and complex images, far past animation. So making digital images should enter schooldays as making digital letters (ASCII!) did very long ago.

I don't want to leave the impression that everything is bad and no progress has been made. It would be absurd to say that in a volume honoring Carl Sagan. We have had a lot of success. We need more.

Success and Failure

The successes are quite plain to everyone by now. They are rarely in the mass media, more often in trade books and toys. One instance: The Klutz Press of Palo Alto publishes remarkable books, many exactly tuned to the wry young teenager, alert, anxious to learn, and yet scornful of pomposity and indifferent to customary duties. They put out books exactly in that idiom that yet contain, not only substantial, serious instructional reading matter (with numbers), but also the materials of experience and experiment bound into the book one after another, an aluminum plate, a battery, a package of rice – so you could cook the average rice ration of the poorer world. These books are highly successful. They sell out printing after printing. They really work very well.

My wife Phylis and I have long reviewed children's science books at Christmas. We cover the trade books in science for children (that means books sold to readers and not mainly issued by schools as text-books) published in the English language, mostly in the United States. We now see about 550 books per year. When we started in the midsix-ties, we saw 300. Of the 300, perhaps 150 were somewhat embarrass-ing, 100 were workaday, under 50 were excellent. It's still true that under 50 are excellent, but there are no more embarrassing books, or at most a handful. They're all workmanlike, they're all worthwhile, any adult beginner in any field, looking at two or three books, will find one to taste as a start on any new subject, Ethiopian grain crops or how CD-ROMs work. This is really quite an improvement.

Another success is highly visible, beyond the world of children and school. Let me allude to The Planetary Society as only one fine example. Theirs is a sophisticated top crust of amateurs of the solar system and its exploration. But I can mention many other groups: the telescope users and would-be telescope users, the radio amateurs, the people who hunt and work rocks and crystals, the automobile rebuilders, above all, the very large group, not quite in science but in a technology both antecedent to and deriving from science, who are gardeners and birders. They incorporate tens of millions of people. Theirs is not quite science, but it is recognizably an effort of extending order by finding new experience.

What we have not made is a wide success in the public schools. That doesn't mean there are no good public schools; there are superb ones. But it means that by and large the challenges to the schools are heavy, and science has not helped as much as it can. I do not think it will be much better in the schools as long as we insist upon translating the participatory and the sensory by promulgating content mainly in words. That is a very difficult thing to do at best, and dangerous to do in the present circumstances. How rapidly it can become a catechism! Many school boards will accept a cheaper book whose authors certify that every sentence of the new standards (that means a few dozen sentences for a year of biology in high school, something like that) has been included.

I don't believe scientists should promulgate unquestioning answers, but rather urgent questions and the material to approach them – that includes of course prior answers. I'm not saying codified knowledge is without justification; it is essential. There are claims for uniformity; there are demands to tighten instruction up. We can't allow things to go on purely by *laissez faire* and so waste children's lives, they say. I weigh such arguments, but I am more afraid of the other side. Life is inert in the frozen state, and so is science.

The Old Disorder

Let me conclude with a remark or two that are badly needed. (Ann Druyan touched upon this subject gently.) Inside the house of science, all is not well. We American scientists are most productive and engaged. But that house doesn't look like America. We're changing it, but too slowly. We don't look like America: For we don't include one-half women, or one-tenth Afro-Americans, or one-tenth Hispanics. There are certainly many reasons; it's not a conspiracy; it's not all culpability; it's not all indifference. It's much more complicated than that, but we have solved many complicated problems. This one we have to solve; until we do, all efforts at a decent education in science in our democracy are endangered. We may well fear a general rejection, a wide disbelief without examination by people who indeed have great need for the empowering gifts of science.

I saved one grand old rock to stand on at the last. Like art and like music, the deep, powerful, beauty and pleasure of science are an accomplishment of our human society, one we are obliged by our human unity to share. I close by referring to Carl Sagan. In the evening lecture of the birthday symposium (also presented in this volume), he demonstrated what a single talk in a single lecture room can do if it combines candor and passion. He was manifestly not blinded by his heartfelt passion. He listened to the questions with care, he made contact by ready paraphrase, again and again he fished out something good from the concerns of young people at the microphones. It was a humane ending to a beautiful lecture – and even though he showed us only one illustration, it was memorable. For your luminous example over many years, Carl, we thank you.

15
The Visual Presentation of Science

JON LOMBERG[*]
Honaunau, Hawaii

Some years ago I was asked to produce a videotape about the Strategic Defense Initiative for the Union of Concerned Scientists. As part of the research for this video, I was directed to speak to Dr. Richard Garwin of IBM. His advice guided the depiction of the weapons systems that would be shown in my animation. There was one sequence in which a Soviet missile successfully penetrated the Star Wars shield high over Cape Cod. Dr. Garwin described to me the appearance and trajectories of the multiple warheads fanning out over New England.

"What color are the warheads?" I asked. He was momentarily taken aback by my question.

"I don't know," he replied at last. "It's so.... so...... so unimportant."

"Not to me," I said, "I have to paint them."

We talked about it for a while and decided they should be white. No detail is too trivial for a scientific illustrator to consider, and if you think the color of the warheads doesn't matter, how would it have affected the image if I'd made them pink or plaid?

Figuring out the right visual question to ask is just one of the many special challenges facing the artist inspired by science.

The artist must of course understand the principles involved in the subject being rendered. For example, the simple tasks of positioning

*Jon Lomberg has collaborated with Carl Sagan since 1972, when he illustrated *The Cosmic Connection*. Since then he has worked with Sagan on many other projects, including most of Sagan's books and the Nuclear Winter studies. He was Design Director for the Voyager Interstellar Record. Lomberg was Chief Artist on Sagan's TV series *Cosmos* and won an Emmy Award for "Outstanding Individual Achievement in Creative Technical Crafts." He was Project Director for the Mars-bound CD ROM "Visions of Mars" and Co-Designer of the Cassini/Huygens Diamond Medallion. His art may be viewed at <http://planetary.org/tps/art.html>.

a ring around a planet correctly or showing the phases of the moon properly are inevitably botched by any artist innocent of the relevant physics. The ringed planets and crescent moons in editorial cartoons are never quite right.

In astronomical art it often seems that half the things you paint are really too faint to see; the other half really too bright to look at. A truly realistic image would result in a black on black or white on white canvas. So you must enhance color, stretch contrast, exaggerate, but exaggerate knowledgeably and convincingly.

Hardest, but also most rewarding, you must work with the scientists, on whom you rely for the raw material of your art. Most scientists think analytically, not visually, and are better at predicting numbers than appearances. The exploration of the solar system has shown how poorly astronomers envisioned the supposedly bland face of Jupiter or the supposedly blue skies of Mars.

If astronomers have not been successful at predicting how things will look, they have more than made up for it by providing real images whose beauty and power to inspire have provided new visual icons for our age in the photographs taken by telescopes and NASA spacecraft, especially the Viking and Voyager missions. One of the great benefits of being alive now has been the chance to see those images.

When it works, the collaboration between visual art and science can produce results that are important for both disciplines. People are sometimes surprised that an artist is inspired by science. But if a sunset can inspire a painting, why not the Sun itself?

The personification of the blending of art and science is of course Leonardo da Vinci (Figure 15.1). However, it is worth remembering that Leonardo predated organized Western science by a century or more. He didn't regard himself as a scientist, but rather as an observer of nature. He was therefore spared the false dichotomy posed

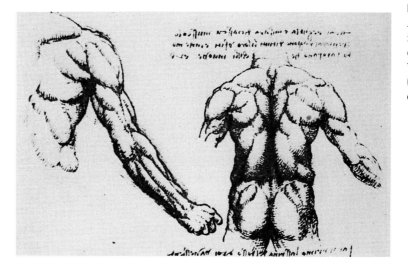

FIGURE 15.1
Anatomy study from the notebooks of Leonardo da Vinci (1542–1619). (Courtesy Library of Congress)

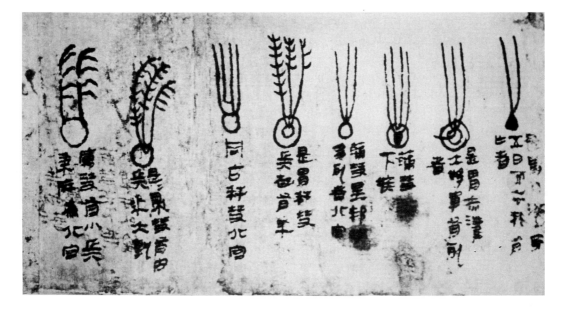

FIGURE 15.2

A portion of the Mawangdui silk, *ca.* 300 B.C. (Courtesy of F. Richard Stephenson)

to all of us in this field, who are frequently asked "Are you a scientist or an artist?" Artists and scientists both observe nature, the one to understand, the other to interpret. Leonardo's consummate blending of both disciplines represents a long tradition of partnership, with a pantheon of great names, especially in biology and medicine, among them Vesalius, Haeckel, and Audubon, in whom scientist and artist were perfectly combined.

The human proclivity to copy nature and thereby document the universe predates Leonardo. In fact, it is ancient and widespread (Figure 15.2). The Mawangdui silk, an extensive Chinese atlas of comets, dates from 300 B.C. but records many, much older observations. This image appears in the book *Comet* that Carl wrote with Ann Druyan to celebrate the return of Halley's Comet in 1986.

Similar recordings of astronomical observations played an important role in the development of astronomy in Europe, such as in the *Cometographia* of Johannes Hevelius of Danzig, who rendered feathery ink drawings of comets that were observed between 1577 and 1652 (Figure 15.3).

There are thousands of other examples in sciences ranging from astronomy to zoology. The realistic rendering of natural phenomena was an important tool of science before the invention of photography. This is art in the service of science but also, science in the service of art because many of the images stand alone for their beauty, even apart from their scientific utility.

It was once thought that the invention of the camera would eliminate the need for the scientific illustrator, but the eye and hand can do many things that the camera cannot, such as eliminating extraneous

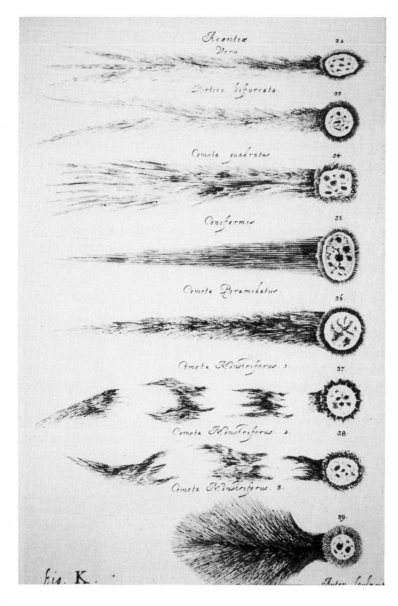

FIGURE 15.3
From
Cometographia by
Johannes Hevelius,
1668. (Courtesy of
Donald
K. Yeomans)

detail and emphasizing certain aspects of an image. Even after the introduction of photography, artists continued working in many fields of science, medicine, and engineering. Traditional media have recently been supplemented by new animation and computer graphics technologies, affording great opportunities both for presentation of data in scientific publications and meetings and for dramatic new popular visualizations, especially in print and broadcast media. Clear and striking illustrations are still required whenever a topic in science must be shown to an audience of nonspecialists.

Carl Sagan's great talent lies in communicating science, and he has always understood the importance of imagery in that endeavor.

Cosmos may be the work by which Carl is best known. The words and images of *Cosmos* have been seen and read all over the world and have had a great influence on people's understanding of science and the universe. As Chief Artist for the series, it was my responsibility to conceive of ways in which many topics in astronomy, physics, and biology might be depicted for a general television audience (Plate XI). Plate XI shows a scene from a sequence we called the "Cosmic Zoom," which was an imaginary trip through the universe in which we approached and entered our Milky Way Galaxy.

The art and animation we created for *Cosmos* had to enhance and support Carl's style of presentation and be, like his words, accurate yet accessible, rigorous, and at the same time stirring.

Visual sequences like the computer interpolation of evolution from cell to human, the dramatized life of Kepler, or the Cosmic Zoom showed how effectively visual media could capture people of all ages and backgrounds.

The *Cosmos* series was one of the first science shows to utilize the new powers of digital imagery. We worked out an arrangement whereby James Blinn and Charles Kohlhase at the computer graphics laboratory of NASA's Jet Propulsion Laboratory developed and produced for us the animation of the earliest Voyager encounters as well as several other sequences in *Cosmos*. These animations were among the first to exploit the power of computer graphics for the visualization of science in imagery sophisticated enough for the national media to use, and reuse, and reuse. They've become icons themselves (Figure 15.4). (In fact, the series *Star Trek: Voyager* not only borrows its name from

FIGURE 15.4
Computer animation by James Blinn, Jet Propulsion Laboratory.

the NASA spacecraft, but one of the opening graphics is an homage to Blinn's famous animation of the Voyager spacecraft's passage through Saturn's ring plane.)

Digital imagery has become increasingly important in the visualization of science. As wonderful a tool as the computer is, it is only as good as the visual skills of the person using it. All power tools are conveniences that can be extremely dangerous when used improperly, and computer graphics are no different. When the technique calls too much attention to itself, the image can overwhelm the subject and the content can be lost. Scientists getting their hands on graphics software can be like novice painters who buy a huge assortment of pretty colors and then are determined to use every one of them in every painting. Just because a graphic technique is available, it doesn't mean that you have to use it. Numerical data presented in three-dimensional contour maps, with exaggerated vertical relief, may look very pretty, but the imagery can actually obscure or distort the truth of the data. A little restraint goes a long way when the image must inform as well as dazzle.

A notorious example of digital misinformation is the widely released animation showing radar maps made by NASA's Magellan spacecraft, converted to a topographical simulation of the Venusian surface. Vertical relief had been greatly exaggerated, and Venus's relatively flat surface was transformed into a completely distorted landscape of dramatic peaks and towering volcanoes. Whatever verbal disclaimers accompanied the animation, most people walked away believing what they saw.

The classical techniques and principles of painting still play an important role in the visual presentation of science, and the skills of the artist determine the effectiveness of a piece of scientific art. In the field of astronomy, images have been supremely important, and artists have been almost as important as the astro-photographers in teaching people with their images. This style is epitomized by Chesley Bonestell, who is considered the consummate master of the genre. His work in books and magazines in the mid-twentieth century was enormously influential and helped inspire the generation of scientists and engineers who began the exploration of the solar system (Figure 15.5).

Bonestell created a complete atlas of the landscapes of the solar system, based on contemporary scientific knowledge of the planets. Of course we have since learned that many of these ideas were wrong. The lunar horizons seen by astronauts look far softer and less interesting than the jagged and dramatic peaks Bonestell rendered in his many exciting paintings showing human explorers on the lunar surface. It has been suggested that one reason there was a sense of anticlimax after the Apollo program was that the real moon was a disappointment after Bonestell!

However, one advantage the artist has over the scientist is that even a painting depicting an obsolete idea can still be valuable as a work of art, such as Bonestell's image of Saturn as seen from Titan. We now

FIGURE 15.5
Saturn seen from
Titan by Chesley
Bonestell.
(Courtesy Space
Art International)

believe that the clouds of Titan would always block a view of Saturn,
and you could never see the planet from the surface. An old scientific
theory proposing a clear atmosphere for Titan is simply wrong and
is completely obsolete and devoid of all but minor historical interest.
But a painting of a theory later proven incorrect doesn't lose its power
as art and continues to inspire wonder and provoke curiosity.

Bonestell's realistic spacescapes have inspired artists in many
countries who continue to paint in the what-it-would-look-like-if-you-
were-there style. Some of us worked together to create the images seen
in *Cosmos*.

But photographic realism is not the only style in which art and science may be usefully combined. Interpretive use of motifs and ideas
from science can be used to convey concepts and ideas as well as

landscapes. One painting from *Comet* illustrates the power a more graphic approach can have (Plate XII). A dinosaur contemplates a spectacular shower of comets in the evening sky sixty-five million years ago, unaware that a comet was about to strike the Earth and wipe out most of the species alive, including all the dinosaurs. The individual depicted had something like hands and a larger brain size for its body weight than most of its contemporaries. If the dinosaurs had not been rendered extinct, perhaps the dominant form of intelligent life on Earth today would have descended from such a creature. I felt that the more stylized, graphic treatment of the scene added a dimension of poignancy that a more realistic version would have lacked.

Some scientific art seems like the visual equivalent of the scientific paper, but sometimes they can be poems too. An anthropologist friend, Dr. Richard Lee of the University of Toronto, who had lived with the !Kung people of the Kalahari, had informed Carl that the !Kung name for the Milky Way translated to The Backbone of Night. When Carl told me this, he was thinking of writing an astronomy textbook with that name, and I was inspired to illustrate this evocative metaphor for him. The phrase and image ended up not as a textbook, but as the title and signature image for one of the episodes of *Cosmos* (Figure 15.6).

The logo for The Planetary Society, the space interest organization that Carl cofounded with Bruce Murray and Louis Friedman,

FIGURE 15.6
"The Backbone of Night" by Jon Lomberg©.

FIGURE 15.7
Logo of The
Planetary
Society©.

combines a sailing ship and the planets (Figure 15.7). This symbolizes
the rebirth of the inquiring and exploratory spirit of the great age of
European naval exploration as our species now begins the exploration
the Solar System. The logo is derived from Carl and Frank Drake's own
depiction of the solar system used on their Pioneer plaque and from
my own drawing, "A Space Caravel," which appeared in Carl's early
book *The Cosmic Connection.* That drawing was itself inspired by
an engraving by Breughel. Images like this set a tone and motivate
popular interest, by emphasizing the romantic aspect of scientific enterprise. Perhaps it is this appeal that has helped make The Planetary
Society so remarkably successful in initiating and supporting projects
in both planetary exploration and SETI, the Search for Extraterrestrial
Intelligence.

It may even be that our ability to express scientific ideas visually is a
key to establishing communication with extraterrestrials. The metallic
phonograph records carried aboard each of the twin Voyager spacecraft bear greetings, music, and a description of our planet in sounds
and images (Figure 15.8). In assembling this sequence of images, the
principles of clarity and simplification that guide the creation of scientific illustration were useful in conceiving of the best way to create
a visual message from Earth. Though most of the images recorded on
the Voyager Record are photographs, we also used techniques such
as molecular diagrams, cutaway views, and overlay transparencies to
present complex information about our planet in such a way that an
intelligent extraterrestrial might understand it.

Frank Drake and I had to devise a way of helping ET understand
the purpose of the Voyager Record. We came up with a diagram which
is engraved on the aluminum container in which the record awaits
discovery among the stars. The diagram on the cover of the record

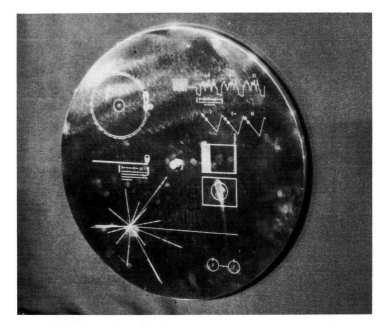

FIGURE 15.8
Voyager record
cover. By
Frank Drake and
Jon Lomberg.

shows how to place the enclosed cartridge on the metallic disk, how fast the record is to be spun, how long it takes to play one side, and how the square wave video signals recorded on the disk are to be reconstructed into images.

Once the picture sequence has been decoded, other graphics attempt to establish an understandable symbol system, such as symbols for the elements associated with diagrams of atoms. Then the indicated elements are shown forming the nucleotide bases of Earth life's genetic alphabet. Similarly conceived diagrams show the planets of the Solar System, structure of the Earth, composition of the atmosphere, and other specific and quantitative facts about our planet. One actual example of scientific illustration appears in a page from a book by Sir Isaac Newton, showing the primary method by which humans store information, where the method of launching an object into Earth orbit is diagrammed (Figure 15.9). Even if extraterrestrial recipients recognize this picture as showing some sort of writing, the text of this page must remain forever enigmatic. However, it is possible that the renderings of orbital trajectories will be understood by an alert alien intelligence.

Some of the principles developed for interstellar messages have recently been pressed into service for a more mundane but urgently important cause – the safe disposal of nuclear garbage (Figure 15.10). In 1991 I served on a panel charged with the problem of marking the proposed WIPP (Waste Isolation Pilot Plant) in New Mexico with a warning message intended to remain effective for at least 10,000 years. This required an interdisciplinary approach combining physical science,

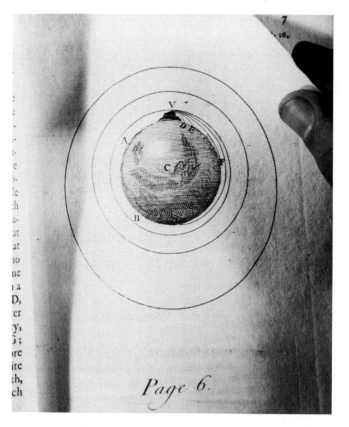

FIGURE 15.9
Page from *System of the World* by Isaac Newton. (Courtesy National Astronomy and Ionosphere Center)

anthropology, and psychology, with architecture and art. We investigated how very durable physical markers might be constructed and how best to record a message cautioning against inadvertent human intrusion into the buried waste repository. Our final design recommendations employed a redundant system of large and small markers, inscribed with markings in written languages, symbols, and pictographs, warning people against digging into the site. In addition to individual markers defining the perimeter of the site, the entire site might be marked by arranging large mounds into symbolic shapes, using material with a different radar brightness than the surrounding desert floor. The site might thereby be made visible to aircraft or spacecraft observations.

Drawing from experience in thinking about SETI issues, our team investigated how symbols and pictographs might be used in markers designed to inform our remote descendants of the hazards of an ancient waste dump. Some of the principles elucidated in the design of messages for extraterrestrials were incorporated, such as the need to begin from first principle in attempting to describe anything. As in SETI, a good message is self-explanatory. If we want to use a symbol like the radiation trefoil, for example, we must include in the message a pictograph explaining the meaning of the symbol. We could not

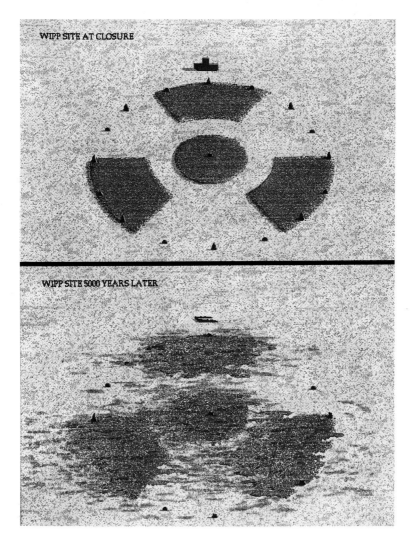

FIGURE 15.10
Design for nuclear waste marker by Jon Lomberg, from "Expert Judgment on Markers to Deter Inadvertent Human Intrusion into the Waste Isolation Pilot Plant," Sandia Report SAND92-1382, 1993.

simply rely on symbols remaining in human memory, their meaning unchanged, for millennia.

There is a certain irony that skills developed for astronomical art could be pressed into useful service on important down-to-Earth issues at the confluence of technology issues and public welfare. The most dramatic example is the use of art in depicting the Nuclear Winter hypothesis. In 1983, Carl was one of a group of scientists who released a study on the effect of dust on the climate. The results were published in the scientific literature, but Carl also had the wisdom and the courage to present this new theory to the general public in the widely circulated newspaper magazine PARADE. In addition to

FIGURE 15.11
Cover for *PARADE*
magazine©,
October 1983.

publishing their work in scientific journals, the group presented these dramatic and important findings to the public at large in *PARADE* magazine (Figure 15.11).

Carl asked me to prepare a videotape that he narrated showing the climatic effects of global nuclear war. The images of the Earth during and after a nuclear exchange were developed from the models of Carl's group, with additional consultation by Dr. Stephen Schneider of the National Center for Atmospheric Research. This videotape was broadcast all over the world, and images from it were reproduced in many periodicals. Giving people images that illustrated this hypothesis played at least some role in stimulating the widespread discussion and debate that followed, by allowing nonscientists to visualize the mechanism that would cause a nuclear winter. The tape was shown to political leaders in the U.S. Congress, the Canadian Parliament, and in many other countries in both superpower blocs. David Lange, then Prime Minister of New Zealand, used these images to explain his refusal to allow U.S. nuclear warships to dock in his country.

Visualizing science is nowhere as important as in teaching. The SETI Institute's Life in the Universe Curriculum project has produced award-winning materials which use art to help interest children in science. One of our efforts was to produce a slide set showing planetary and biological evolution. Large-scale and close-up views of our planet and its inhabitants at different times are used in conjunction with workbooks and teacher materials to teach this difficult and complex subject. The subject is taught several times in different grades, with different emphasis each time the material is presented (Plate XIII).

Our project has created science teaching guides for grades 4 through 9 and been tested nationwide. We have found that the use of strong imagery is a valuable component of teaching materials. Students of all ages respond to visual materials like these. The text must vary with the age level, but the same images can be used with equal success, which makes the production of such complex visualizations extremely cost effective.

Carl's work has always stressed the necessity of humans adopting a cosmic perspective, a precept I learned from him that has remained at the heart of all my work.

In 1991, the staff of the National Air and Space Museum discussed with me the possibilities for a mural of the Milky Way for their "Where Next Columbus" gallery, then under construction. I proposed that we make this an image that was not schematic, like my galaxy painting from *Cosmos*, but truly cartographic, that we plot the location of the best-known galactic objects in the best map of spiral arm structure. To my surprise, this had never been attempted before. Working with Jeff Goldstein of the Museum's Laboratory for Astrophysics and with Leo Blitz of the University of Maryland, I spent almost two years researching and painting this piece, a 6 × 8 foot canvas now on display in the Museum (Plate XIV). Imagine an X connecting the diagonally opposite corners of the artwork. Our Sun is at the center of the X, about two-thirds of the way out from the bright and glowing galactic center. We are in a small bridge of material called the Orion Spur that connects the Sagittarius and Perseus spiral arms. Most of the stars you can see with the naked eye are in a very small circle right around the Sun (about the size of a small coin on the original canvas). What we call the Milky Way in the night sky is only a view of the nearer arms. Most of the galaxy, and nearly all the light from the center, is entirely blocked by clouds of dark, opaque dust. Included in the artwork are miniportraits of the Orion, Trifid, Lagoon, North America Nebulae, Eta Carinae Nebulae, and Eagle Nebulae, as well as the globular cluster Omega Centauri, and almost 300 other known clusters, nebulae, and other objects. (A more detailed view of this art is visible on a large poster of this painting that is available through The Planetary Society in Pasadena.)

As in maps of old, the farther you get from Earth, the more conjectural the fine detail becomes. Nevertheless, this work remains the most accurate image of our galaxy yet produced by human beings. It will be a very long time before we gain a true view of the detailed structure of our galaxy. Perhaps the most thrilling aspect of painting this work was feeling that I was the first person ever to see how the galaxy might actually look.

(And some science may have come out of this piece as well. Dr. Goldstein noticed that our projection showed an asymmetry in the distribution of globular clusters around the galactic center, an asymmetry that seems to line up with a faint barlike structure in the nucleus that Dr. Blitz had discovered, depicted for the first time in this painting. Dr. Goldstein is now writing a paper on this evidence of a possible coupling between the clusters and the nucleus.)

I wish I could say that the fruitful collaboration Carl and I have enjoyed for over twenty years is typical of the way scientists work with artists. But Carl's support of and reliance on art has been the exception rather than the rule among scientists. I believe that the importance of images in communication has largely been undervalued and underutilized by the science community. I am sorry to say that much of this stems from a kind of intellectual arrogance that marginalizes the skills of art and design as somehow less difficult or specialized than the skills of science. Over and over again, the artist is brought into projects only after all the conceptual decisions about materials and presentation have already been made. It's as if the contribution of artists is limited to filling in small details of execution, when the only question left to resolve is the color of the warheads. Usually, the naive way scientists envision an image, a poster, or an animation shows their lack of experience in this area. By neglecting to bring visual experts into the process at the formative stage, scientists often miss opportunities for effective presentation.

Once I attended a meeting NASA convened about improving its communications. Sitting around the table with the NASA Administrator were many bureaucrats, a dozen scientists, a few teachers, a few science writers, but except for me and one other artist/astronomer there was nobody expert in the visual presentation of science: no book or museum designer, no magazine art director, no filmmaker or photographer, no television producer, no special effects wizard, no computer graphics genius. Just looking at the roster in attendance made it clear what NASA's problem in communication is. And it's a problem by no means restricted to NASA.

Expecting the scientist to create or even conceive of the way something can be communicated visually is like expecting the visual artist to engineer the scientific instrument. Great scientists who are also great writers are so rare that we justly celebrate the work of Carl Sagan and the handful of others who can both do the science and explain it well. Rarer still is the scientist who can do the visualizations unassisted, when really top-notch graphics are required. Even the most common venues of communication, the classroom or public lecture, could be improved if the scientist had been required to take a semester course covering such topics as slide management and presentation.

Our world is becoming more visual, less verbal. The generation of scientific illiterates about whom we profess such concern demand images if they are to read or listen to the words. And these images must be strong enough to compete with the blizzard of advertising and pop culture images that bombard us each day. Otherwise, people will simply not pay attention.

Few enough universities offer courses in the fields of science writing. Almost none teaches scientific illustration, and the few that do concentrate on botanical and medical renderings rather than physics, chemistry, or space sciences. Many times over the years I have been approached by young artists who want to know how to pursue a career like mine. I wish I knew what to tell them. There is no career trajectory for such artists. The vast scientific establishment in this country basically ignores the discipline. There is not a single institution of higher learning in the United States that offers training in the kind of work discussed here.

Scientists sometimes deplore powerful images as having only "gee-whiz" value, as if emotional appeal were somehow inconsistent with the transfer of information. Yet art, in all its varied manifestations, hi-tech and low, is still one of the best tools humans have invented to communicate. There is a long and honorable history of scientific illustration, complete with its heroes and martyrs. One such is Sydney Parkinson, who sailed in 1768 with Captain James Cook and Sir Joseph Banks in the *Endeavour* on a great voyage of scientific discovery. Parkinson was the expedition's artist, "a gifted and amiable young man" according to Banks, and he drew everything from the plankton Sir Joseph hauled up from the Pacific to the aboriginal inhabitants of New Zealand. After surviving the many arduous rigors of the voyage, including shipwreck on the Great Barrier Reef, Parkinson succumbed to a fever brought aboard ship in Batavia, where *Endeavour* had stopped on its way home. Yet the 743 plates based on Parkinson's drawings that form the core of *Banks' Florilegium*, provide an enduring legacy worth far more to science (and to Banks) than the paltry seaman's wages Banks thought appropriate recompense to Parkinson's surviving family in England. Parkinson's family had to wage a protracted lawsuit against Banks to recover more suitable compensation. Banks, one of the wealthiest men in England, did not stint on the cost of producing his book, and later in life was more generous to other artists in his employ, perhaps partially out of guilt at his rather shabby treatment of Parkinson. The glory of Parkinson's contribution compared with his worldly success is a fitting symbol of the ambivalent status the artist of science has always had.

The role of the artist in the communication of science is an important part of the tradition of science, and we would be unwise to neglect it now. Carl Sagan has always been one of the great friends and champions of scientific art, and in this as in many other areas, his example is a good one to follow.

16
Science and the Press

WALTER ANDERSON
Parade Publications

It is a pleasure and an honor for me to participate in the celebration of Carl's sixtieth birthday.

If anybody should be talking about science and the press, it should be Carl himself, because in this century only a handful of people – like Carl, Isaac Asimov, among others – have brought scientific comprehension to so huge an audience in so elegant and authoritative a fashion.

Because I'm the one who has been asked, I'd like to briefly give you some idea of the current state and significance of science journalism, at least as it seems to us on the editorial print side of the fence. First, though, I want to describe another characteristic that Carl and I share:

We're kind of like the little snail on the tree in early spring, and there we start working. And high up in the tree, a couple of birds are looking down, and they're teasing the snail: "Where are you going? What are you doing?" And the snail looks up and says, "This is a cherry tree." The birds keep teasing and saying, "But there are no cherries on it." The snail says, "There will be by the time I get there."

Carl and I kind of stick to it.

I'm amazed at how scientists lose their discipline when it comes to the art of communication. If anybody believes that a magic hand put its little finger on Carl's neck and made him a fine writer, then you don't understand writing or communicating. That's simply not true. I can always tell an amateur because an amateur comes up to me and says, "Mr. Anderson, I love to write." No professional writers love to write. They hate to write. They love to have written. It's hard work.

You scientists have to learn how to communicate better. You should be visiting your arts courses, your communications professors. Improve yourselves and discipline yourselves. If you have 10 minutes, speak for 10 minutes. The audience likes that.

I suppose there was once a time, not so many centuries ago, when science news wasn't covered at all by the popular press. There are two reasons for this: First of all, science didn't really seem to affect the everyday lives of most people – or, if it did, people didn't really know about it. Second, there was no popular press. Back in the sixteenth century, the great Danish astronomer Tycho Brahe built his own print shop so he could disseminate his heavenly observations, and Galileo Galilei didn't exactly call a press conference to announce that the Earth still moved, regardless of what the Pope said. But all that has changed, and basically in our own lifetimes.

Today, science news makes up an astonishing proportion of the daily news flow. This hasn't happened because editors have suddenly decided that science news is good for you. Sure, most editors do like to contribute to the public well-being and to publish news that should help the readers. But no publication will stay alive, and no editor will stay employed, unless people read his or her product. So the great recent upsurge in scientific reportage, not only in specialized publications but also in large-circulation newspapers and magazines, reflects an urgent public demand for, and insistence on, scientific news.

As a result, there is a new phrase that is being heard increasingly, both in educational and journalistic circles: science literacy. It means an ability not just to read about science but also to understand it. To be scientifically literate, you have to know more than that the eustachian canal isn't a great waterway or that supercollider doesn't mean a bad driver. You can't just learn the words; you have to acquire a comprehension of how science operates and to understand both the promise and the perils that technology holds out to the world.

When I was growing up, about five hours from here, at the edge of the Bronx, just two publications seemed to offer the possibility of scientific enlightenment to me. They were called *Popular Science* and *Popular Mechanics*. Both tried to explain the latest technological advances in practical rather than philosophical terms. In preparing for this talk, I looked at the science section of a reference work called *Magazines for Libraries*. I discovered that nearly fifty current publications were listed, ranging from *Omni* and *Discover*, which were described as "intended for the casual browser," to *Scientific American* and *Technology Review*, said to be aimed at the more serious reader. I'm much more concerned with the way science is handled not in scientific journals but in the general, everyday press – that is, in daily newspapers and in mass-market publications like my own.

Soon after I became editor of *PARADE* in 1980, I shared some time with William Shawn, then the editor of the *New Yorker*. Later, after Carl had done a particular article for us, I reviewed it with Mr. Shawn. I wanted to see his response to it, and he really was quite impressed. He said, "Walter, as far as readers' understanding is concerned, you have *Scientific American*, the *New Yorker,* and then *PARADE*. What do you think about that?" I said, "Well, in order to understand *Scientific American* and the *New Yorker*, you have to read *PARADE*."

PARADE has a circulation of 37 million. What does that really mean? Simmons Research tells us that it translates into more than 80 million readers. *PARADE* is not only the largest-circulation publication in the United States, it is in fact the largest-circulation publication in the world, ever. Who better than Carl Sagan to teach us about science in *PARADE*? It comes out every week, it goes to every state in the union. Its audience includes everyone from truck drivers and waitresses to professors and politicians.

Like all nonspecialized publications – and that includes daily newspapers up to and including *The New York Times* – we face the challenge of writing about science in a way that will be informative and enlightening, but also interesting and stimulating, for readers whose curiosity may be only fleeting or superficial, as well as for those who have a keen interest in black holes, the Big Bang, or the demise of the dinosaurs, but who are deficient in technological background. This is a tough thing to do, particularly in daily newspapers, which by their very nature must be put out in a hurry.

Ben Bradlee once referred to journalism as "history on the run," which is probably the best description I've heard. Good reporters are always at a premium in this business, and good reporters who can write in clear and graceful English about a specialized subject – whether it is science or music, architecture or ice hockey – are even rarer.

As I'm sure you know – or, if you don't, someone has told you – the press does occasionally make mistakes. I think one of the most common and most serious mistakes is that of perspective. I'll include all the press in this. One of the most common examples is the golf ball and grapefruit form of science journalism: "We went to the Moon, viewers. Here's the grapefruit – that's the Earth. There's the golf ball – that's the Moon.... Now we can all understand it." Actually, if we were going to do this correctly, the Moon would be over by that wall, and that's what's really impressive.

In 1910, when Halley's Comet came through, the Earth was to pass through the tail of Halley's Comet, and gases would be released. Well, that's accurate, and on Earth, of course, what happened? Worldwide panic. The Pope issued directives. People committed suicide. Now, if you took that whole tail, you could fit it into a briefcase. Nothing bad happened. Perspective. Perspective.

I'll give you another example. When you get falling-down drunk, 80,000–100,000 brain cells are lost. Just think about that. You get falling-down drunk one night, and you lose 80,000–100,000 brain cells. And that's reported in the press. I read it in *Time*, in *Newsweek*, in *The New York Times*. It's pretty serious. Of course, if you drank every night, it would take you 1200 years before you destroyed your brain cells. This is not to encourage alcoholism or drinking but to tell you that, if you want to have fun from time to time, you can!

Scientists are particularly prone to detect what they regard as errors, though these may seem trivial or even picayune to a harried editor. But occasionally, though it pains me to say this, it is the scientist rather than the journalist who is responsible for misleading or inaccurate information in the press. Most researchers are content to publish their findings in appropriate scientific journals, but there are some, unlike Galileo, who have held press conferences to announce breakthroughs. I'm sure that many of you will recall a famous incident of a few years ago regarding two scientists at a university in Utah who announced – prematurely, as it turned out – that they had achieved fusion in a glass jar at room temperature.

More recently, there has been a battle, fought at least partially in press releases, between French and American scientists claiming primacy for an important discovery in AIDS research. So both scientists and science writers have to develop a certain mutual caution and mutual respect in keeping the public accurately and honestly informed of the true significance and import of scientific discoveries. Sometimes we both get help from unexpected sources. The tragic, brutal slayings of Nicole Brown Simpson and Ronald Goldman did more to call DNA to public attention than a dozen science expositions.

In the important but imperfect world of science reporting, it is vital to have writers who are trusted by people on both sides of the fence, people who are the leaders of the science community and people who just read the papers. That's what makes Carl Sagan unique, and not only as a scientist who can write. He is a human being deeply concerned with what's happening on our planet, not to mention a few other planets. Just a few of the titles of his *PARADE* articles will serve to show the range of his concerns: "Star Wars: The Leaky Shield" and "The Question of Abortion," written with his gifted partner, Ann Druyan. We received more than 330,000 phone calls following that abortion article. The editor of *PARADE* – you know him – modestly said: "It is the finest article written on that subject ever." It certainly was read more than any article written on that subject, ever.

The first time that I visited Ithaca was to spend some time with Carl Sagan, and it was at a time when he had just faced death. Many of you know of or remember the experience. He was really weak, and I was in the living room with Annie, and Carl came out, and he walked very slowly. He sat down, and he began talking to me about something of concern to him – a project that he was working on with 100 other scientists around the world. He was so weak, he only stayed a few minutes, and I said, "Why don't you write that for *PARADE*? Why don't you tell everybody about it?"

He went back to his room, and later, when I left Ithaca, I was amazed that here was this man who had just faced death, and his concern was the fate of others. He did write the article for *PARADE*. It was called "Nuclear Winter." It created quite a stir among thoughtful people throughout the world. If ever there was a scholar with the properly human touch, it is Carl Sagan.

I also have to admit that one reason I am so proud to be associated with Carl is that he is one of the foremost exemplars of the power of the written word. I'm giving away no secrets when I acknowledge to you that our friends in television also play a part in enhancing scientific knowledge and building interest in new technologies. But being what is known in the modern jargon as a print journalist, I like to think that it is through words – as well as, if not even more so than, images – that the basic messages of science can best be conveyed.

None of us will ever forget the thrill we experienced in 1969 when we saw those first televised images of the men walking on the Moon. Yet, as the years unroll before us, these images will surely be effaced by even more spectacular and revealing photos of space exploration. It's happening now. But the words, the words spoken by Neil Armstrong, the first man on the Moon – "That's one small step for man, one giant leap for mankind" – will never be forgotten.

Today, as in Galileo's day, we live by words, and even as we try to advance science literacy, we must remember that literacy in general is a problem that also needs to be urgently addressed in this country. It would be impossible for me to give a talk on any subject at any time and not mention literacy. I don't apologize for it; it's the crux of my life.

In my country, our country, 44 million Americans – roughly one in five – function at the most marginal level of literacy, according to the U.S. Department of Education. So here too is a challenge that must be faced by all of us who write and who read. Let me emphasize that I believe science and technology are going to play an even more important role in the press tomorrow than they do now, simply because science and technology are going to play an ever increasing role in our everyday lives. If there are medical advances, we want to know about them. If there are environmental problems, we want to be warned of them.

Remember the radon scare of a few years ago? Nobody ever heard of radon a few decades ago – plus, in the press, it was a word that looked like a typographical error. Yet it was through the press that its dangers were ultimately made clear and corrective action was taken in so many areas. Yes, we should have been more alert to the significance of the AIDS plague, but medical authorities were also slow to size it up.

I'll take advantage of the moment to suggest a frontier I'm particularly interested in. Since the hand can't grasp itself, and the eye can't see itself, I wonder: Can we understand consciousness? And I look forward to the scientific exploration of consciousness, whether it is with microbes or larger and more complex organisms.

I've had great teachers in my life. Three of them, quickly:

• One was a gunnery sergeant, Sgt. Shimkonas. When I was made a Marine sergeant in Vietnam, I said to him, "How can I motivate the troops, Gunny?" And he said, "Andy, grab 'em by their private parts, and their hearts and minds will follow." I think he meant I should get their attention.

- I once asked my mentor, Elie Wiesel, "Does my life, does any life, have value?" And he answered, "In and of itself, no. It is up to us to give it value." The existential response.
- And what does Carl say about our responsibility? If we begin by saying, "I am responsible," we cannot only change our lives, we can also build a better world. Who is responsible? I am responsible.

Carl has taught me how important it is to live for a more noble motive. Your life – each of us – is worth a noble motive.

The men and women who work to improve the quality of life, to educate our children, to safeguard our natural resources, to probe the future of our universe have every right to expect a comprehending but not necessarily cooperative press. Remember, we need the sand fleas and the ticks.

The Sea of Galilee and the Dead Sea are both fed from the same source – the moisture from Mount Hermon – and yet they're so different. Compared to the Sea of Galilee, the Dead Sea is truly dead. It has far less life, it doesn't move, it's not appealing or appetizing. The Sea of Galilee is rich with life. Why? It has an outlet. It empties on to the Jordan plain, which it enriches. People are like that: Some people are like the Dead Sea. They get to get. Others are like the Sea of Galilee. They get to give.

In the future as in the past, it will be the task and the challenge of scientists and newspeople to travel together in mutual understanding along the information superhighway. I hate that word. I don't know why I wrote it. It's the dumbest thing. I hear Vice President Gore say, "We're on a superhighway." What does that mean, superhighway? Well, I suppose it's going to happen, and it will be called something else.

We're going to have to walk in mutual understanding along the information superhighway and whatever other roads lead ahead. As far as *PARADE* magazine is concerned, I hope that my friend Carl Sagan, a man who truly gets to give, will be around for years to make that journey with us.

17
Science and Teaching

BILL G. ALDRIDGE

Science Education Solutions
(Former Executive Director,
National Science Teachers Association)

Science teaching in the United States faces serious challenges at all levels. As our technology has expanded almost exponentially in diversity and utilization among all sectors of our society, the knowledge base associated with that technology has also expanded, often to increase even further the number of subdisciplines of science or of engineering. Accompanying such changes have been a variety of societal and global problems that have technological or scientific components.

In the face of such dramatic change, teaching in our schools, and indeed in most of our colleges and universities, remains (in stark contrast to the training facilities among U.S. corporations) largely tied to a third-world technology: textbooks, conventional laboratory work – when and if a laboratory exists – and a teacher standing near a chalkboard and lecturing to a group of students sitting in chairs. The prominent and well-publicized exceptions to this scenario are confined to a very small number of schools in which vastly greater resources have been placed than in the overwhelming majority of other schools.

The response by many educators to our scientific, technological, and societal challenges is to broaden, but not deepen, the science learning experiences of their students. In addition, there is a concerted effort on the part of many prominent science educators in schools of education at colleges and universities, individuals with large followings among science teachers, to respond to technological, scientific, and societal challenges by configuring science learning to a highly speculative model fashioned from a mix of postmodernist philosophy and something called radical constructivism, which denies the kind of objective reality that is inherent in the replication of measurements and observations by independent scientists and which denies the universality of the laws of nature [von Glasersfeld 1992].

Science teaching also faces three other serious challenges: first and foremost, and almost unique to the United States, the strongly pervasive belief among students, parents, and, indeed, among scientists, even in the face of contradictory evidence, that real science can be learned only by a few people with certain special and inherent abilities and aptitudes and that the majority of us can only learn something quite superficial and often characterized as science literacy; second, the inability of most science educators, and the failure of most scientists – who do know the differences – to make clear distinctions for students and for the public among observations and empirical data, or relationships and empirical laws, those scientific results that are subject to replication, and the theories and models created by the human mind and used to explain or account for those empirical aspects of science; third, the blurring of distinctions between science and technology.

The third-world character of most schools is a direct consequence of our failure to provide adequate resources to those schools. The prominent and well-publicized exceptions to this situation are confined to a very small number of schools in which vastly greater resources are being placed than in the huge majority of other schools. These exceptional schools fall into two categories, schools for the advantaged and a few schools for the less advantaged, often enhanced to such levels so as to blunt public criticism over the sad state of other similar schools having meager resources. The exceptions, schools like the Illinois Academy, the North Carolina School for Science and Math, or the Thomas Jefferson High School in Virginia, spend at least ten times more per student than is available to other schools. For the most part such schools do little more than skim off the most highly motivated or highly advantaged students from other schools, along with a disproportionate share of resources. They create a mechanism for providing additional advantage to the advantaged, a good example of the biblical Matthew Effect.

The average high school in the United States spends about $6 per year per student on materials, supplies, and equipment for science. Given that there are about twelve million students, of which only about 25% are studying science in a given year, we currently spend some $24 million per year. To upgrade the school technology would require an increase of at least two orders of magnitude in the level of that kind of support, increasing the cost to about $2.4 billion per year. It is therefore especially offensive to hear, as we often do, the false assertion that teachers resist technology in the schools, and that is why they remain with their third-world technology.

The thrust to broaden science education in schools takes on many forms. The rationale is simple, but logically defective. It goes something like this. Since the world is becoming technologically and scientifically more complex, and we face myriad global and societal challenges, science education must assume the same level of complexity in the sense that it becomes not just interdisciplinary within science, but that it becomes multidisciplinary, extending to include health, economics, civics, and social studies.

Because many problems in these areas have scientific or technical components, it is argued that students should learn the science in those contexts (which, according to a National Research Council report [Druckman 1994] does not provide for transfer of learning). The alternative of learning fundamental concepts, principles, and laws of science that pervade all such applications is shunned on the grounds that it represents traditional science education, which, by most measures, we would agree, has been a failure.

The logical fallacy has two parts. First, just because the world is complex and our global problems cover many areas of science and nonscience, it does not follow that the methods of science can solve those problems or that learning to use science words and terms in their contexts leads to any ability to deal with such problems. The other fallacy is in the conclusion that the failure of traditional education in the fundamentals of science requires this particular new alternative (the fallacy of the excluded middle). There are other alternatives by which essentially all students can learn the fundamental concepts, principles, and laws of science and can use them in applications now and, more importantly, in the future world and in its new and different technologies and its new and different set of global problems.

The most serious issue in science education is the latest fad sweeping the nation: radical constructivism and the postmodernist view. Whether wrongly interpreted and used or not, there is a major push to create *constructivist* teachers and to inject this element into state science education frameworks. Its rejection of the objective character of science and the assertion that all knowledge is created subjectively and individually offers fertile ground for those who want students to create their own individual knowledge of the natural world, regardless of its match to reality. Radical constructivism as used in science education is a peculiar relativism of science knowledge, and it is a perversion of the common wisdom that when you teach someone something, you must start with what they already know. If such science educators knew the distinctions between empirical science and the theories or models we create to account for that empirical knowledge, their focus could legitimately be on helping students modify their preconceptions to better match those scientific models and theories scientists have created and tested.

There are, within radical constructivism and postmodernism, elements of anti-Western science, if not antiscience. Gerald Holton has addressed this issue quite comprehensively in his book, *Science and Anti-Science*.

The problems associated with constructivism are directly connected to the ability to distinguish empirical science and theoretical science or to distinguish among elements of science, like concepts, definitions, empirical relationships, or theories and models. When we call light by the name light waves, we are imposing the wave model of light onto the phenomenon. In fact, there are light phenomena that require a photon theory. Our failure to reconcile particle and wave models is not a paradox of nature. It is our failure to find a more comprehensive model or theory. A fine example of such a distinction between empirical science and theory has been provided most eloquently by Stephen Jay Gould in *Natural History*, where he distinguishes between the facts of evolution and the theory of natural selection. The empirical components reflect what happens in nature; the theories represent the ability of humans to create explanation. The latter offers opportunity for alternatives – each to be tested for its predictability. When our use of terms or language imposes a model or theory on natural phenomena, we remove from consideration student explanations of those phenomena. We take away the most exciting and interesting aspects of science, coming up with testable hypotheses and creating models or theories they can compare with the body of empirical evidence.

Making distinctions between science and technology is really seeing the difference between the small number of scientific concepts, principles, and laws that account for essentially all natural phenomena and all technical or engineering applications or seeing science as this enormous body of fact and information that surely must be impossible to comprehend. To emphasize too strongly applications to the present technology at the expense of far more lasting fundamentals deprives the student of the ability to adapt in his or her understanding of a changing technology.

What about inherent ability and aptitude for science? Sure, there are exceptionally talented people like Carl Sagan. But they represent a tiny minority who will likely learn regardless of their circumstances. The majority of us are far more alike in such abilities than we are different. And there is considerable evidence that all of us can learn science with depth of understanding, if we are expected to do so and we are given the opportunity and resources to do it.

Now what is the National Science Teachers Association (NSTA) trying to do to help students learn the fundamentals of real science and its applications? Through support from the National Science Foundation in a multiyear national project, NSTA is trying to reform secondary school science [Aldridge 1995; Aldridge, et al. 1997]. The project is called, *Scope, Sequence and Coordination* (SS&C).

The idea for this reform was conceived in the early 1980s, primarily in response to practices in U.S. secondary schools that filter out the majority of students from higher level courses in science. In what has been characterized as a layer-cake curriculum, the majority of U.S. students study biology in grade 10, less than half go on to chemistry in grade 11, and only about one in five persist to take physics in grade 12.

The present layer-cake curriculum has serious defects. Few students complete the sequence, producing very small percentages of graduates who have studied the physical sciences; individual courses are sequenced according to the logic of the discipline and not in terms of levels of abstraction (e.g., in grade 10 biology, the periodic table, with quantum numbers, is presented as a basis for molecular and structural organic chemistry; but these topics are simply a concentrated form of what must be learned a year later in chemistry); science materials or instruction fail to take into account student metaphors and preconceptions; terminology is introduced and then explained, when what is needed is experience before terminology; having separate subjects results in students failing to see any relationships among science disciplines or applications to the real world; single, 180-day courses fail to take into account the benefits of spacing, which allows for repetition at successively higher levels of abstraction.

The layer-cake curriculum, with its intrinsic limitations and weaknesses, can be shown to be one of the leading causes of poor performance of U.S. students relative to that of students in other industrialized nations, where each science discipline is studied every year in the secondary grades. The widespread practice of tracking in the United States also places many students at a disadvantage in pursuing the study of science, whether or not their ultimate goals are in fields of science.

There is clear evidence that science opportunities at the high school level in traditional courses are very limiting for minority students. These students are "...more often inappropriately placed in dead-end general education or vocational programs, are provided fewer resources, the worst teachers, and are often deprived of hands-on experiences" [Oakes 1990]. Evidence shows that very large numbers of extraordinarily talented young people are not identified under the present layer cake, tracking system because they are filtered out at an earlier stage.

Recent data show that the layer-cake curriculum precludes science and engineering careers for some 79% of African-American, 78% of Hispanic-American, and 78% Native-American youth who entered college in the fall of 1993, compared with 28% of Asian-Americans and 67% of White or other racial or ethnic groups. These are the numbers of entering freshman who have not studied the essential prerequisites of three years of science in high school (physics, chemistry, and biology). Had elements of scope, sequence, and coordination of secondary school science (SS&C) been in place in their schools, there would have been opportunities to pursue college careers in science and engineering for at least 980,000 more young people who entered college in 1993, among which were at least 195,000 more African-American, Hispanic-American, and Native-American youth.

Our goal is to replace the layer-cake curriculum for the few with a program in grades 7–12 in which essentially all students learn science every year in four subjects: physics, chemistry, biology, and the earth and space sciences. This newly designed program is sequenced appropriately, and the science disciplines are coordinated in such a way as to help students see the relationship among the sciences as well as the applications of science. Most importantly, the project is developing materials and methods to achieve the new National Science Education Standards being created through the National Research Council of the National Academy of Sciences. Widespread implementation of SS&C will increase the pool of talented students, while also shifting its makeup to be more representative of the diversity of our population in terms of currently underrepresented groups.

What are the essential elements of the NSTA's reform in science education? The SS&C reform has as its fundamental tenets the following:

1. Provide the learning of science in four subject areas each year, biology, chemistry, physics, and the earth and space sciences;
2. Explicitly take into account students' prior knowledge and experience as expressed in their preconceptions and metaphors (much of this is available in the literature);
3. Sequence content, and the learning of it, from concrete experience and descriptive expression to abstract symbolism and quantitative expression;
4. Provide concrete experience with science phenomena before the use of terminology that describes or represents those phenomena;
5. Revisit concepts, principles, and theories at successively higher levels of abstraction;
6. Coordinate learning in the four science disciplines so as to interrelate basic concepts and principles;
7. Utilize the short-term motivational power of relevance by connecting the science learned to subject areas outside of science (such as history, art, and music), to the practical applications of how devices in our technology work, and to the challenge of solving those personal and societal problems that have relevant, underlying scientific components;
8. Utilize the long-term motivational power of sudden and profound understandings of science and of the awe that stems from comprehension of the power and universality of a relatively small number of fundamental principles of science;
9. Greatly reduce topical coverage, with an increased emphasis on greater depth of understanding of those fewer fundamental topics;
10. Create assessment methods, items and instruments to measure student skills, knowledge, understandings, and attitudes, both for program evaluation and for the requirement of assigning grades, which are fully consistent with tenets 1–9.

The NSTA reform of secondary school science education has completed its middle school phase and is in its high school phase. Some 13 schools are in the initial group of schools, including one from New York State. Schools include the full diversity found in our nation, including schools that are almost entirely Hispanic, African-American, Native American (this, a boarding school, has some 450 students from 65 different tribes), or Asian. They are also at every socioeconomic level. These schools, which will produce the first graduates who can achieve the National Science Education Standards, will be followed by others that may begin implementation in the fall of 1996. Thanks to Microsoft,* ninth and tenth grade student and teacher materials are distributed free via the Internet: http://www.Gsh.org/NSTA_SS and C.

For sustained reform of science education, teachers must learn science well and in several fields, colleges and universities must reform what and how they teach science, and our expectations and attitudes of who can learn science must change. And scientists must accept responsibility for defining what science is and what science is most important to learn.

BIBLIOGRAPHY

Oakes, J. 1990. Multiplying Inequalities. *The Effects of Race, Social Class, and Tracking on Opportunities to Learn Mathematics and Science.* Santa Monica, CA: RAND.

von Glasersfeld, E. 1992. Questions and Answers about Radical Constructivism. In Pearsall, M. K., ed., *Scope, Sequence and Coordination of Secondary School Science: Vol. II. Relevant Research.* Washington, D.C.: National Science Teachers Association.

Druckman, D. and Bjork, R., Editors, *Learning Remembering Believing: Enhancing Human Performance* (National Academy Press, Washington D.C., 1994) pp. 25–56.

Aldridge, B., "High School Science Reform: Taking SS&C to a higher level," *The Science Teacher* (NSTA, Arlington, VA, Oct. 1995) pp. 38–41.

Aldridge, B., Lawrenz, F., and Huffman, D., "Scope, Sequence and Coordination: Tracking the success of an innovative reform project," *The Science Teacher* (NSTA, January 1997), Vol. 64, #1, pp. 21–25.

*Microsoft Corporation has supported the global school house site, and they are also using SS&C materials on their Encarta site: http://encarta.MSN.com/SCHOOLHOUSE

SCIENCE, ENVIRONMENT, AND PUBLIC POLICY

18
The Relationship of Science and Power

RICHARD L. GARWIN

IBM Research Division

I am delighted that Carl Sagan has arrived at his sixtieth birthday, that he has accomplished so much that such a book is more than appropriate, and that I have been invited to provide a paper on "Science and Power."

Science (the knowledge of the nature and function of the world and its parts) connotes power, if only sometimes the power to know when to get out of the way.

Intervention involves more often technology as well as (or instead of) science, and technology evolved for a long time independent of formal science. Now, of course, the advance of science has made it much more relevant even to older technology and essential to modern technology.

The encounter of scientists with military or political power has not always been pleasant. Whereas the death of Archimedes was an unsought consequence of war, Galileo's recantation was a victory of dogma over freedom of speech, if not freedom of inquiry.

Science provides power both absolute and relative – relative, that is, to the power of someone else. Absolute benefit may allow one's society to improve the quality of crops, to learn the nature of the planets. Relative benefit may be more immediately valuable, as in the tale of the two hunters, George and Mike, pursued by an enraged grizzly bear. Running as if his life depended on it, George after awhile called to Mike "I don't know why we're running, everyone knows you can't outrun a grizzly bear." And Mike replied, "I don't need to outrun a grizzly; I only need to outrun you."

So, while rulers may have sought a court scientist for novelty or for enlightenment, and eventually for practical contributions, it was the relative benefit, especially in military activities, that forged the closest links of science to temporal power.

That brings us to science in the service of competition, which is not the same as competitive science. Some scientists are competitive by nature – a quality that tends to be prized in modern life and that, with necessary regulation, underlies a lot of the wealth creation and advances of modern times. It is sometimes said about the competition inherent in horse racing that its purpose is to improve the breed and I suppose that is true; however, one can also win a horse race by drugging the opponent's horse or bribing the jockey.

One of the major forms of competition is war, with which science has had a long association. During World War II, scientists in Britain were motivated by patriotism and fear of conquest to see what they could contribute against the Nazi threat, and, spurred by refugee scientists, the United States mobilized its scientific community to produce the proximity fuse, to help develop and manufacture radar, and to create the atomic bomb – the first two nuclear weapons used against Japan in August 1945.

War is hardly a game, particularly when accompanied by a plan and a program for genocide. The goal of the Allies was both laudable and necessary – to stop the Nazi war machine and soon also the Japanese military. The means, though, was to destroy enemy fighters and equipment and, eventually, industrial support and people.

Through individual genius and genius of organization, combined with dedication and energy, this crucial battle was won. But science and technology were used effectively on the other side as well, notably in long-range rockets.

After the war, the United States was left with an enormous facility for producing weapons and what was seen to be an enormous science-based system for inventing and developing them. Aside from the personal tragedies and combat deaths, the United States did not suffer materially from the war and so was not faced with the immediate enormous task of reconstruction, as were its allies Britain and the Soviet Union, and the defeated or liberated powers like Germany and Poland.

In the immediate postwar atmosphere, there was no obvious military threat, but there was momentum created in the wartime laboratory efforts, together with the excitement and substantial government support. Most of the scientists left weapons work for university activities, with science now to receive substantial funding from a grateful government and people. Conventional industry, for the most part, really did not know what to do with science. American industry, however, had unmet consumer needs and went back to making automobiles, refrigerators, rail cars, and light bulbs.

A surprising amount of initiative was required, as reported by Simon Ramo, particularly, eventually to create a totally new science-based industry for modern weapons, their command and control, and intelligence.

But it was clear that if large-scale conflict should come, the United States would be better served by more advanced (even by less costly) versions of the weapons that had been so important during the war, and Los Alamos turned rather slowly to making improved versions of fission weapons, to introducing the concept of the boosted fission bomb, and eventually solved the problem of a practical approach to a thermonuclear weapon. From concept in early 1951 to ten-megaton explosion in less than twenty months was no sluggish program.

The advance of miniaturization of vacuum tubes and then the burgeoning of semiconductor electronics after the invention of the transistor at Bell Telephone Laboratories allowed the practical realization of enormous amounts of computing and control capacity within the weapons themselves and facilitated communication where such computation needed to be done off-board, so to speak. So the wartime advances in propulsion, structures, and particularly in industrial organization were followed by successive generations of weapons and weapon systems such as those devoted to air defense, integration of platform and weapons, and the like.

But at the same time that this enormous peacetime weapons industry arose, there was a potential enemy, even identified (and perhaps partially created) by some far-seeing individuals (or paranoid) during World War II itself. This was our only possible rival at the time – the Soviet Union. And Stalin was a formidable foe, ruthless with his own people. His organized terror was accompanied by personal terror, with the result that people feared to approach him to argue vigorously against activities that were harmful to the Soviet Union and even to the system that Stalin was trying to create. The destruction of Soviet biology by Stalin's elevation of Lysenko may have been prevented in physics only by Stalin's need for the physicists to create the Soviet atomic bomb.

About these matters, we have now a good deal of information, much of it from people whom we have grown to know quite well, such as Roald Sagdeev, Georgi Arbatov, and many others. In regard to the Soviet atomic bomb, we have now the scholarly book by David Holloway.

But on our side, we had no such individual terror that would act against people who spoke frankly to our presidents. Some did not have the opportunity, but all too often, those who did have the opportunity were unwilling (for what seemed to them good reason) to provide advice that might have helped. What are these reasons?

First, I suppose that there is self-doubt, although this is not high on the list of infirmities that one would ascribe to many of those in a position to talk with presidents.

Second, there is the desire to preserve one's influence for the future and not to sacrifice it on something that might be a lost cause. Perhaps a little more about this later.

If advice is secret, who knows what goes unsaid? But with regard to public advice to those in power, I don't remember a time when there has not been criticism from one side or the other, or more commonly from both. Several partial solutions have been achieved. First, one can try to have as advisors working scientists who bring with them the honesty and self-questioning that are essential to successful science. Furthermore, the peculiarly American mobility of individuals among the roles of outside expert and provider of congressional testimony, full-time government employee, and consultant is helpful, and we have had some success in spreading this to other countries.

The vast majority of scientists active and effective in public policy are based in universities, and universities in this way play a vital role in our democratic system. Cornell, Massachusetts Institute of Technology, Stanford, and Harvard have been among the universities from which the most effective contributors have come.

Beyond the universities, nonprofit public interest groups like the Federation of American Scientists (FAS), having its fiftieth anniversary in 1995, play an important role, largely in conjunction with academic scientists. You have read an article by an official of FAS, Ann Druyan, and you are reading one by another here, since I am Vice-Chairman of the FAS and Chairman of the FAS Fund.

Also particularly valuable in government service are people like Spurgeon Keeny, now President of the Arms Control Association, and the late James R. Killian, first head of the White House President's Science Advisory Committee (PSAC). These typify people who are not professional scientists but who have the integrity and the combination of confidence and self-questioning that are essential to science.

Accompanying scientific influence on important matters is the temptation to manipulate others. There is also the use of power and influence against the individual scientists, their colleagues, or even their institutions or families.

For instance, at a time when Edward Teller was pushing hard for a commitment to the hydrogen bomb, J. Robert Oppenheimer, as head of the General Advisory Committee of the Atomic Energy Commission (and fabled wartime Director of Los Alamos) stood in his way. It is only natural that one should try to accomplish one's goal, and also to remove the obstacles, and that is apparently what Teller tried to do in this instance.[1]

When Hans Bethe and I published in March 1968 our *Scientific American* article, "ABM Systems," the Secretary of the Army, Stanley Resor, signed a memo asking the Army to marshal support among scientists for the system that was threatened by our arguments.[2]

---

[1] I add that the advance of civilization derives in part by self-limitation from the natural.

[2] Stan Resor is now a staunch ally in the fight for rational, even real, defense, programs.

As an aside, from my own experience the more difficult problem for those interested in substance is not to counter individuals who are knowledgeable and committed on the other side, or individuals who are committed though ignorant, but rather to counter paid publicists, or legislators, or those who regard it as their job to be hired guns and to do whatever is not clearly illegal to further the goals of those who are paying them. In 1991, Ted Postol of Massachusetts Institute of Technology had the ingenuity to analyze television video of putative intercepts of Iraqi Scud missiles by Patriot air-defense interceptors in Israel and Saudi Arabia, and the courage to publish these results earned him organized attack by Raytheon, the Patriot system builder. The definition of a successful intercept now seems to be that an incoming missile was detected and an interceptor successfully launched.

The Strategic Defense Initiative (SDI) program initiated by President Ronald Reagan on March 23, 1983, with a television broadcast that surprised not only scientists outside his Administration but also the scientists and military inside, is instructive in the relation of science to power.

The Executive Summary of the SDI study led by James Fletcher in 1983, following (not preceding) President Reagan's announcement of the SDI program, did not fairly represent the contents and conclusions of the seven volumes of the study. When asked, Fletcher publicly acknowledged having had no influence on the Executive Summary and when asked who wrote it said, in my hearing, "Beats me. Someone in the White House, I suppose." But he did not contest publicly the substance of the summary. Programmatically, SDI dissected the necessary technological advances into manageable pieces that could be parceled out as contracts to industry – each one a reasonable or major extension of our capability. However, to reach the goals of SDI would have required success in a vast number of these elementary improvements, as well as the cooperation of our adversary, the Soviet Union – both inherently unlikely.

A Director of SDI, Lieutenant General James A. Abrahamson, was not appointed until almost a year after the Reagan speech. Both before that time and afterward, a major influence was played by Major (now Colonel) Simon P. Worden, an astrophysicist with whom many of us had vigorous and often unpleasant encounters in our analysis of prospects of success of SDI. Later, he apologized to some of us for his actions as self-acknowledged hired gun, but that did not help us or the nation at the time.

Unfortunately, it is rare for a hired gun to do what is required even in the most noncontroversial scientific field, and that is to provide a reasoned paper – not just a viewgraph. But I did have an extended correspondence with Peter Worden following his claim that a 10-meter diameter mirror in low Earth orbit could be used to focus sunlight to cause damage on the Earth's surface, as if there were not a totally fundamental difference in this case between the disorganized light from the sun and that from a laser.

It is more than a quip that it is not so much what you don't know that will hurt you but what you think you know that isn't so. In this regard, I have repeatedly admired Carl Sagan's dedication to challenging his own tentative conclusions. I was not so pleased when a well-known scientist from the Lawrence Livermore National Laboratory, active in SDI activities, had not assimilated by 1985 a simple analysis of 1983 that emphasized a single point: If an offensive missile can be destroyed in its boost phase of four minutes' duration, by a very fast interceptor that needs a launch weight 100 times its payload to reach the required speed, then it will require an interceptor launch mass of 100×100 or 10,000 times its payload to destroy a missile with a two-minute boost phase.[3] It is irresponsible for scientists or others who are playing a role in advocacy not to know of the chief arguments of the other side.

Now consider three cases:

- If with gun in hand, I accost a prosperous-looking person on the street and demand, "Give me $100 or I'll kill you" and I am caught, I will be sent to jail for armed robbery.
- If without the gun in my pocket, I accost the same person and say, "Give me $100 or my brother will kill you" and I am caught, I will go to jail for extortion.
- But if I go on television and demand from the public, "Give me $300 billion for our military activities, or the Russians will kill you" I will be deemed a great patriot and perhaps will be elected to high office.

There are real hazards and opportunities in this world, and that is why we cannot always err on the side of caution in response to every claimed threat. Furthermore, matters of arms and the military must be considered together with possible perceptions and responses by other nations and the stability of the interactive system.

Arms control and disarmament are important options that we are finally beginning to use.

What scientists can do to help our country and our world is very much limited by the inefficiencies of our current political process. For the activities of some of those in public office, the outcry of attorney Joseph Welch to Senator Joseph P. McCarthy in the 1954 hearings of Senator McCarthy on alleged communist influence on the U.S. army is appropriate: "*Have you no shame?*"

---

[3]This simple analysis was essential to showing that pop-up interceptors have no future for defense against a responsive missile force – that is, one that takes into account the nature of the defense.

His personal power challenged in the academic porkbarrel activities, Representative John P. Murtha (D-PA), Chairman of the House Appropriations Defense Subcommittee, lashed out at Representative George Brown (D-CA), Chair of the House Science Committee, by deleting $900 million in academic research sponsored by the Department of Defense. Indeed, *Aviation Week and Space Technology* of October 3, 1994, has an editorial, "Abolish the R&D Porkbarrel."[4]

And while many, but not all, in the House and Senate posture for the electorate and admittedly spend the majority of their effort on amassing funds and credits for reelection, the staff of the two elected officials in the Executive Branch of the U.S. government have been doing the same for several decades. Our Legislative branch seems to be 90% posture and 10% performance; Common Cause (no surprise) has not achieved its goals of reform. Paradoxically, it may be that the United States could have a more coherent long-term policy if officials were limited to a single term in office, so that they could concentrate on doing the job to which they were elected.

The rare combination of outstanding scientific talent and dedication to the public interest so apparent among the invited guests of the Sagan sixtieth birthday symposium will be of no avail if our society cannot govern itself in those matters to which science is not central. Democracy, which we prize, contains the seeds of its own destruction. The power to choose includes the power to choose wrong.

In Russia there is a hazard of rejection of democracy, but also in the United States. Unless we provide more effectively the public goods of security against crime, of employment, and health care, I see a real threat of the electorate choosing remedies that will lead to disaster. If those of us who have some power don't address these problems because it is right to do so, we should do it because our future depends on their solution.

[4]The following week, the House–Senate conference committee reduced the cut to $200 M – about 10% of Department of Defense-sponsored university research.

19
Nuclear-Free World?

GEORGI ARBATOV

Institute of U.S. and Canadian Studies,
Russian Academy of Sciences

We are witnessing an interesting paradox. As soon as nuclear weapons appeared, there was (at least officially) an almost unanimous consensus that these weapons should be outlawed, banned, and destroyed. A vivid demonstration of it was the very first resolution of the United Nations' General Assembly, unanimously adopted in January 1946. America in those years introduced proposals intended to do away with nuclear weapons. I have in mind the famous Baruch plan. To ban the atomic bomb was also the position of the Soviet Union. The differences that prevented an early agreement were mostly about details, not principles (so it seemed, at least). Reporting on the moods of that time, the late Phillip Noel-Baker said: "Not a single voice was raised in any country against the proposal that atomic weapons should be eliminated from national armaments; the press of every shade of opinion was unanimous."

In the fifty years that have passed since nuclear weapons were tested for the first time and a short while later used, we learned a lot about them, about their military and political values, including the basic, I would say, ultimate fact, agreed upon by the two biggest nuclear powers much later: that a nuclear war cannot be won and must never be fought. Though the terrible consequences of use of nuclear weapons were more or less clear from the very beginning, the way to this conclusion (I do not even speak here about its implementation in the practical structure, doctrine, arsenals of armaments of the armed forces) was not an easy one. And in early post-Hiroshima days some military and political leaders more than once seriously considered the possibility of using these weapons, arranged military exercises with their use, and then sometimes made conclusions that, to use the words of then Minister of Defense of the USSR, N. Bulganin, the nuclear weapons are not as terrible "as it is said by the imperialists," which in his opinion, among other things, meant that experience gained during

228

the last war could be used in the future with only slight corrections (*Nezavisimaja gazeta* 6.06.1991).

The understanding that one cannot use these weapons in war came only very slowly, and this conclusion was preceded by many attempts to devise a technical or strategic gimmick, which would make it possible and safe to use them. Antiballistic Missile Star-Wars was only one of them.

Strangely enough, this understanding did not impede the nuclear arms race. The latter, it seemed, acquired a life of its own, and the more obvious it became that nuclear weapons cannot be used for any rational goal, the more intensive became the quantitative and qualitative race in these weapons and means of their delivery, sometimes even justified by their uselessness as a means to wage war. One of the gimmicks to make them look usable was smaller size and selectiveness (for instance, the so-called neutron bomb intended to kill only people and leave intact cities, buildings, plants, equipment, etc.).

Promises of real change have appeared only with the end of the Cold War. But the way to it was long and arduous, and the question still remains whether this change is irreversible.

During this fifty years the whole geopolitical situation on the globe had to change radically. It was by necessity a long and painful process. The world went through quite a few serious political crises and local, though very destructive and bloody, wars. At times it seemed that a nuclear conflagration was just around the corner. For decades everything that accompanied the Cold War, indeed constituted its substance, flourished: the arms race, subversive activities, slanderous propaganda, and other instruments intended to defeat or at least to destabilize each other.

It seems that only after the major powers have tried out everything intended to win the Cold War without success, did they come to a conclusion that aside from big dangers and fantastic expenditures, such relations promise nothing. The Cold War, like the nuclear one, cannot be won and therefore it would be much better for both sides to end it. Each country then can get down to its real business – that is, to her own domestic economic, social, and political problems – which to a large degree were neglected because of the Cold War and therefore have become more and more complicated and, in time, dangerous.

This is, in my opinion, why and how the Cold War at last came to an end. Of course, aside from the obvious futility of the efforts, and increasing exhaustion, there was also a need for that famous boy from a fairy tale, who would say that the king has no clothes on. In other words, there was a need for a nation and a statesman who would have the courage and skill to initiate the process and show the folly of the whole exercise, thereby depriving his partner in Cold War games of an enemy, and bringing it almost automatically to an end. This essential role was played mainly by the USSR, by M. Gorbachev and a few of his colleagues.

The end of the Cold War radically changed the relations between the former mortal enemies, who were also major nuclear rivals: the USSR, as well as the heir of its nuclear arsenal, Russia, and the United States. Even if they had not yet become full-fledged friends, a military conflict between them, especially a nuclear one, is already simply unthinkable, as well as a military conflict between Russia and any other Western country.

And – here is the paradox – despite all these important changes, the chances to create a nuclear-free world look to the public today even more distant and vague than half a century ago (this does not mean that in reality fifty years ago the plans and hopes to do it were not an illusion).

At least these problems appeared again among those that are being actively discussed. I would refer in this connection to a Pugwash monograph, "A Nuclear-Weapon-Free World," 1993, and "Nuclear Weapons," by E. Bahr (Ditchley Park, 1993).

The situation looks strange indeed. The nuclear powers have come to a conclusion that these weapons cannot be used in a war, that they are, so to speak, political and psychological weapons, intended to prevent an aggression by the threat of a retaliation that will inflict unacceptable damage. But now also the threat of an aggression has disappeared, which makes nuclear weapons redundant even as a political weapon. The door to a nuclear-free world seems open. At the same time, however, this goal seems not less, maybe even more, distant than it seemed half a century ago.

What are the major reasons for it?

Why does the idea of a world free of nuclear weapons sound even in the present, radically improved political conditions, though well intended, still idealistic, based on illusions, and not practical?

Listening to the opponents of the idea one cannot get rid of the impression that maybe, as it was feared by the authors of *Dr. Strangelove*, we indeed learned to live with the nuclear bomb, and even if not to love it, to feel rather well and comfortable in its company. Concerning this fact one can also hear some explanations, among them a few that even sound rational.

Mankind, it is said, for instance, has lived with nuclear weapons for half a century. At the same time, we were spared the experience of nuclear war and even a new world war. Was it not exactly the tremendous destructive power of these weapons that prevented the governments from their use? If so, shouldn't we reject the whole idea of a nuclear-free world as dangerous and obsolete?

This is all the more so because the Cold War has come to an end and the major nuclear powers have normalized their relations. They do not threaten each other with a war, they will decrease radically their nuclear arsenals, and as it looks now they are even ready to cooperate with each other in this field. In such a situation a reasonably small amount of nuclear weapons can even stabilize peace, serve as a guarantee against criminal behavior of some governments, who, having obtained even a few weapons in one way or another could otherwise blackmail the rest of the world.

These arguments deserve, in my opinion, an answer. Yes, the decision to use the destructive power of the weapons no doubt was available to the political leadership of the nuclear powers, and this also meant that getting involved in crises that can get out of control and unleash a big war was much more difficult. But even today we can hardly say for sure that it was due to a wise and prudent policy or to sheer luck that we avoided such outcomes. This was only the dawn of the nuclear era. Can the existence of mankind forever depend on luck and the common sense of political leaders? Among them we have already seen personalities like Hitler and, at least at the present stage of the world's political civilization, nobody can yet feel immune from new historical accidents of that kind. In addition, the chances of irresponsible personalities ruling a nuclear power and making political decisions about the use of nuclear weapons is in a very direct relation to the number of such powers.

Here we come to the heart of the problem. When nuclear weapons continue to exist and remain in the arsenals of even a few selected powers, proliferation becomes practically unavoidable. This fact will radically increase all the risks.

If one justifies nuclear weapons as instruments of deterrence that can prevent aggression and ensure security of his own country, one cannot deprive other countries of a right to enjoy the same kind of security. In case you continue to recognize equality, equal rights for security as a principle of relations between nations, a basic principle of international law, a universal proliferation would ensue.

One more comment: the destructive force of weapons as a factor of restraint for political leadership has a rather relative value. Where is the threshold that will inhibit politicians from their – let it be only an illusion – limited use? There are quite a few examples when even bright people turned out to be wrong in their judgements about the destructiveness of weapons as means to end wars. Alfred Nobel, for instance, was sure that his invention, dynamite, would play such a role. Friedrich Engels, who, whatever one's attitude towards Marxism, was an outstanding social scientist and devoted much attention to military affairs, was sure that the invention of the machine gun will make new wars impossible. Both of them, as many others who made similar forecasts, turned out to be wrong.

One thing, of course, might excuse people like A. Nobel or F. Engels. Their attitude was rational: With the new means of mass destruction and mass annihilation provided by the achievements of science and technology, war was becoming a luxury that mankind could no longer afford. But these rational people underestimated the force of the irrational that is hidden in nationalism: greed, xenophobia, political fanaticism, and missionary zeal, as well as the irrational consequences of an internal struggle for power, which too often influenced, even determined the foreign policy decisions of governments. Who can guarantee that such mistakes, underestimating the consequences of the use of new weapons, would not be made in the future?

I mention these examples to also show among other things that the invention of nuclear weapons was not an accident, but a logical stage in the evolution of the high art of mass murder and mass destruction. The ultimate, terminate weapon could also become something else: a new chemical or biological substance if it had to come from the family of weapons that in one or another form already exist, as well as something more exotic, which is still to be found only in science-fiction novels.

My major point is that in the course of history, technology in general, military technology in particular, as a rule developed much faster and in a much more dynamic manner than the civilized organization of society and of international relations. This created an ever-widening gap, which in turn has opened a rather realistic prospect of a collective suicide for mankind, of a war that would once and for all destroy human civilization.

It seems that the first time people started to see and understand this was during World War I. That is why it was declared "a war which had to end all wars," and the war itself ended not only with the usual, in such cases, act of capitulation of the loser and later with a number of peace treaties, but also with an attempt to create a system of international security under the guidance of the League of Nations, which it was hoped would prevent any repetition of the tragic experience.

This attempt, as is known, failed. World War II was an even more sinister warning about the possible ultimate disaster for humanity. This time the warning was even more difficult to ignore because the war ended with Hiroshima and Nagasaki.

As it soon turned out, this warning was also to a great degree ignored. The Cold War started very soon. And it was inseparably connected with the nuclear threat. We were lucky: The threat, sometimes very great, did not materialize. After the Cold War ended, we finally acquired a unique opportunity to get rid of this threat.

Here I return to the question already raised in the beginning: Why do we not seem to be eager to seize this opportunity? The reason already mentioned is serious, but not the only one. The nuclear era has given life to its own mythology. One of the myths we already touched upon: the allegedly stabilizing effect of the destructive force of nuclear weapons. But this is not the only one.

One of the others is that nuclear weapons are cheap, inexpensive compared to the conventional ones. It is based on oversimplifications and therefore wrong. From our own past we know too well that very soon after the nukes appear in the arsenals, the nuclear powers begin to understand that they cannot be used in any military conflict short of a total nuclear holocaust. Therefore, to have options and to be ready to meet more probable military challenges, one needs a complete set of the most modern conventional weapons, and in big quantities, because the sheer existence of nuclear weapons has militarized the world, and created strong incentives to produce as many arms as possible and even more ... The stubborn facts show that never had the major powers spent more on armament in peacetime than after they became nuclear.

There are many other reasons circulated that explain why it is better to retain nuclear weapons. I, personally, consider only one of them to be really serious. That is the concern whether nuclear deterrence will not again become necessary, due to political instability, doubts whether the present international situation will prevail, and whether it can even develop further in a positive direction.

I think this can become a self-fulfilling prophecy.

The first reason I see here is the heavy burden of militarism inherited from the Cold War. Mountains of armaments, big armed forces, enormous defense industry (in my country about 50% of the industry as a whole) and science, serving the military machine and paid by it, tremendous economic and political interests, which comprise an influential lobby – all of it creates a force, that, in case a convenient opportunity opens up, can turn back the whole course of events (at least for a certain time). I would say more: One cannot exclude the possibility that this force will be able, in the absence of genuine reasons for such turns, to create them artificially.

The legacy of the Cold War should not be reduced to arms, armed forces, and defense industry. It contains also a certain kind of political mentality that breeds suspicion, nationalism, and great power ambitions that easily could lead to serious troubles, political frictions, and even conflicts.

Until we liberate ourselves from this legacy, it is difficult to expect the nuclear powers to become ready and willing to give up their nuclear arsenals, but such developments have not yet occurred in the international situation. One could say more: It is still difficult in the present situation to consider the positive changes that happened in world affairs, in particular the end of the Cold War, to be already irreversible.

I, for one, could imagine a couple of scenarios according to which the developments in Russia will make a return to hostile relations possible. This could be facilitated or even initiated by certain behavior or actions by the United States.

As to my country, I have in mind in particular two partially inter-connected issues: economic collapse and an explosion of nationalism and neoimperialism, followed by the introduction in one way or an-other of a dictatorship. The prospects of an economic collapse have regretfully become rather real as a logical result of the so-called Gaidar reform – shock therapy, which was already tested in many countries (mostly developing) and failed practically everywhere. For Russia these four years of the new economic policy have become a real dis-aster.

Galloping inflation (prices increased by several thousand times); a more than 50% decrease in production; the impoverishment of the majority of the population and the striking prosperity of a few in-dulging in conspicuous consumption; a tremendous increase of crime and corruption, a deplorable state; the decline of grossly underfunded education, culture, science, and health; the deintellectualization and moral degradation of a significant part of the population; a rise of mor-tality and decline of the birth rate: these are the major results of the Gaidar reform. And we have not yet reached the bottom. This means that if the economic policy is not changed in the nearest possible fu-ture, the economy will fall apart. Its collapse will become inevitable.

This will also have far-reaching consequences for our foreign pol-icy. A very probable result of an economic collapse will be, as was already mentioned, a dictatorship (the economic hardships have al-ready increased the influence of extremists from the right and the left, which the December elections of 1993 have shown in a most obvious way). To legitimize itself, to justify political oppression, it will need an enemy abroad (in case of need, one could be invented), and interna-tional tensions. The more so, that in the minds of very many Russians the shock therapy is connected with the influence, advice, and even pressure (political as well as financial) of the West.

The more obvious the failure of Gaidar's (or International Monetary Fund's) reform, the more popular becomes among Russians a theory that it is in reality a conspiracy, that we face here a conscious effort of the Americans and their allies to impose on Russia a policy that would undermine our economy, deindustrialize it, transform us into a sort of Third-World country, the economic function of which will be reduced to a role of a supplier of the West with raw materials and a dumping place for its toxic waste.

I tried to argue many times against such views, saying that though one can find Americans who would welcome such policy, thereby once and for all doing away with the arch-enemy, there are many more Americans who understand that economic chaos in Russia will be followed in this highly militarized country that possesses a great arsenal of nuclear weapons by a political chaos very dangerous for the West as well. But the stubborn support by the Americans of the Gaidar reform makes these reasons less and less persuasive for the Russian ear.

To be honest, such a situation has already started to revive some old feelings of distrust towards the West (in particular anti-American feelings) and together with economic misery a mood of national humiliation, which always breeds nationalism.

Of course, nationalism does not have only economic roots. There is (and not only among the Russians) a feeling of certain nostalgia for the old, big Soviet Union. Many of them feel that they do not enjoy their former security, and that they are not dealt with by other countries, including America, as equals, with proper esteem and respect for their rights. There is also the very real problem of the twenty-five million Russians who suddenly found themselves abroad, sometimes feeling (in some cases justified), discriminated against and limited in their human rights. There are also citizens of other former republics of the USSR, who now live outside of the country to which they ethnically belong.

It is natural that in a former empire nationalism is connected with imperial ambitions, which makes it even more dangerous.

I spoke about my own country, but the problem has a universal importance. It is a problem of each and every country, because the security of the world community and guarantees against the threat of a nuclear war are at stake.

It is a universal problem also in another sense. After the end of the Cold War, it seems, all countries have become too complacent, too sure that the problems of their security were already resolved and they could concentrate their whole attention and resources on other things. I have no doubts that many of them are really urgent. But it is still too dangerous to forget, or to neglect security and foreign policy, especially if one takes into account that Russia, the Russian political elite, because of lack of experience, education, and very often elementary abilities (a lot of new people filled in important posts, without any selection, very often by chance), is hardly ready for bold and imaginative initiatives. We, by the way, have delivered our share of them under Gorbachev, who played quite a significant role in the positive changes of the last years, putting an end to the Cold War. So now it is the turn of Americans, of the West.

Here I want to return to the topic of the nuclear-free world. In my judgement the major weakness of most of the publications devoted to this topic is that the authors treat the problem of complete annihilation of nuclear weapons primarily as a technical one. I understand very well the importance of technical aspects of the problem. If they are not properly cared about they can create an almost insurmountable stumbling block. But the same is even more true when we deal with the political problems that can become an obstacle to a nuclear-free world.

To get rid of nuclear weapons, to ban them, and to enforce effectively a worldwide nuclear-free regime demands a real revolution in international relations, in the approach of the entire world community and first of all of the great powers to the basic principles of their international behavior, to sovereignty, to the use of force, and to fair play in foreign policy.

It should start with a complete stop of the nuclear arms race. Of course the intensity of this race has decreased, perhaps almost come to a standstill. Why do I say almost? First of all, because I am not absolutely sure that the production of all nuclear weapons has stopped. Though we, on both sides, destroy older warheads, one cannot be sure that the production of new weapons has been canceled. The pretext of modernization can justify it, as well as work on new models in the laboratories. The major danger is that the best and maybe the biggest part of the scientific and industrial infrastructure for production of nuclear weapons remains intact. This means that the nuclear arms race can easily resume, at least in theory. I do not think that at the moment it is a real danger.

The end of the Cold War has presented us with a unique chance to end the arms race and maybe to move successfully toward a nuclear-free world. This will of course demand a revolution in our attitudes, our policy, our relations, that I already mentioned.

This revolution as an only alternative to collective suicide is anyway imminent. Sooner or later, because of the threat of nuclear weapons or because of some other achievements of military science and technology, mankind, if it does not change its mode of behavior, is doomed to self-destruction.

Very often, arguing for the noble cause of liberation of our planet from nuclear weapons, specialists try to show the possibility of coming to an agreement to do away with an already invented means of war, pointing to the example of chemical and biological weapons. Sharing their desire to get rid of the nukes as soon as possible, I have to say that this analogy is not quite correct and therefore misleading. If the great powers, without whose consent and even desire the treaties banning chemical and biological weapons could not be concluded, did not possess nuclear weapons, such a ban would hardly be probable.

Maybe exactly the possession of the ultimate weapon made them ready to agree to limitations on less important kinds of means of war, especially because it is much easier for a lot of other, among them smaller and poorer, countries to acquire these weapons, endangering also the great powers, undermining to some degree their monopoly on military deterrence.

To agree to the ban on nuclear weapons would be much more difficult for them, even after the obstacles mentioned are removed. They would expect quite a number of guarantees: guarantees of a state of world affairs that would exclude any military threats to them and guarantees that nobody could cheat, either in hiding a small part of their old nuclear arsenals or in creating clandestinely a few new weapons. And this means full, absolute transparency and most intrusive means of verification. It means also a new role of international organizations, guaranteeing security – an undisputed right and an effective mechanism to enforce its decisions whether on inspection or on destruction of certain installations. In other words, a transition from peacekeeping to peace enforcement by international organizations and under their oversight.

This means a rather long, complicated, and maybe painful way – a real revolution in international relations. But, I am afraid this is the only way that can guarantee our survival.

Disinvention of nuclear weapons is no doubt impossible. The only substitution can be if their ban and complete destruction is accompanied with the invention of a new kind of international relations, which would make these, like most of the other weapons, redundant, useless, and even ridiculous.

Meanwhile, there is a very important and urgent task to prevent theft of weapons or fissionable materials and other ingredients that can enable an international outlaw to create perhaps even a few weapons, enabling him to blackmail the world community. The former Soviet Union is at the moment in such disorder that in this sense it creates particular concerns. I do not want to belittle our special responsibility, but here the need of a broad and very close international cooperation is of exceptional importance.

Now – at least we say so – the former USSR and the United States do not regard each other as enemies, do not feel threatened by each other. Both countries regard nuclear proliferation as the major outside threat. This calls for close cooperation. This also is the main reason why a nuclear-free world is so important for our security.

Understanding the difficulties, some specialists propose compromises, like finite deterrence or a small nuclear arsenal belonging to the United Nations. I think that as important steps to the final goal they might be reasonable and welcome. But the final goal has to remain the same nuclear-free world, free completely, without any compromises, because otherwise the danger of proliferation, which means also danger of a nuclear war, will remain with us, to explode sooner or later.

It is also important to recognize that all the steps to a nuclear-free world correspond to the logic of normal development of international relations, ensuring the survival and well-being of mankind. Considerations of security, economic progress, ecological safety – all point in the same direction. It is not a problem of sacrifices, but of common interests at all stages of this long and difficult way – from the first to the last.

Returning to the present international situation, I have to say that it creates an impression of great euphoria since the end of the Cold War. Yes, this is a great achievement, but a lot of problems remain unsolved. And one has the impression that most of them are not even on the agenda of present negotiations.

To be honest, all governments (and, I think, most of the specialists) were not too prepared for the new, post-Cold War situation, sometimes did not even imagine the problems and challenges that will face us after the Cold War ends or, at least, were not sure how to deal with them. This is one of the reasons why none of our countries has today a comprehensive, really thought-through policy, adjusted to the new realities. In particular, despite serious and, in my country, a desperate economic situation, we spend too much on weapons and armed forces, justifying this sometimes with the explanation that we simply do not have money for disarmament and conversion of military industry. The changes in policy are also too slow and superficial. We, as a rule, react to events and challenges and are too slow with a long-term program aimed at creating a sane and safe international environment for ourselves, our children, and their children and grandchildren. This situation, in my view, is not tenable.

20
Carl Sagan
and Nuclear Winter

RICHARD P. TURCO

*Department of Atmospheric Sciences,
and Institute of Geophysics and
Planetary Physics, University of
California,
Los Angeles*

There was no greater force behind the development of the Nuclear Winter theory, and its application to critical issues of nuclear weapons policy, than the single-minded determination of Carl Sagan. Who else could have had the vision to foresee, even in studying the cosmos, the eventual self-annihilation of our species by weapons of unprecedented destructive power wielded by brilliant reptilian minds? No one else had the breadth of interest and knowledge, nor the dedication and guts to take on the two most powerful and entrenched bureaucratic organizations in history – the defense establishments of the United States and the (ex-) Soviet Union. Nuclear Winter was discovered at the convergence of two widely different paths of thought. One involved Carl's broad interests in life, intelligence, and the origin of the universe. These interests eventually led Carl to the swampy frontier of nuclear weapons policy and physics, into which only specialists dared to venture before. Fortunately, Carl's interests are so diverse that connections between planetary science, the climate, biology, evolution, economics, and public policy become at once logical and natural. In this regard, Carl is unique among his peers. He is not only a first-rate analyst of events occurring in the natural world, he is a scholar of letters with formidable writing skills and, importantly, he is a man of conscience willing to engage the most tenacious dogma. In this light, it is not surprising that he should be found, armed with the penetrating insight of science, wading into the bog of nuclear politics searching for dragons to slay.

The second convergent path to Nuclear Winter involves a series of independent revelations concerning meteor impacts and dinosaur extinction, and a most timely discovery by two atmospheric scientists – Paul Crutzen and John Birks – that smoke might be a problem after multiple nuclear detonations. The latter work, published in the Swedish journal, *Ambio*, in 1982 [Crutzen and Birks 1982] triggered research that yielded the Nuclear Winter hypothesis and set in motion one of the most heated scientific debates of the century. Indeed, soon to follow in 1983 was the seminal *Science* magazine article that defined for the first time a Nuclear Winter [Turco et al. 1983; also Turco et al. 1984, Turco et al. 1990, Turco et al. 1991]. The publication of this unusually provocative and countervailing concept was largely the result of Carl's scientific patronage and dedication.

Nuclear Winter, born at the confluence of these two paths, drove a lance straight into the heart of a firmly established nuclear weapons infrastructure that had been for four decades essentially unconstrained in its assumptions, actions, and budgets. With the publication in 1990 of a detailed analysis of the science and policy implications of the Nuclear Winter hypothesis [Sagan and Turco 1990], Carl finally blazed a rational trail toward eliminating nuclear weapons.

As it happened, the Soviet Union was in a state of terminal collapse at the time, partly under the weight of an impossible economic system, but also perhaps exacerbated by the stresses of international nuclear policies that were at once costly and now potentially self-destructive. In fact, the 1990s eventually witnessed the dramatic restructuring of global superpower relationships, policies, and arsenals. However, despite these unexpected and dramatic developments, the nuclear infrastructure today remains rigid – isolated from the reality of a new world order by trenchant fear and self-interest.

You do not have to be a rocket scientist (or even a physicist) to understand the awesome threat of nuclear weapons. On the desert near Alamogordo, New Mexico, on July 16, 1945, the first manmade nuclear detonation produced a penetrating light that humbled most of the scientists who witnessed it, converting many of them into lifelong campaigners against their own brainchild. Even so, the subsequent nuclear annihilation of the Japanese cities Hiroshima and Nagasaki in 1945 could not overcome prevailing U.S. fears of Soviet armies (and vice versa). The nuclear weapons development programs in both countries expanded exponentially. Trillions of dollars and rubles have been spent since then, and the result is a collection of global nuclear weapons that, even after recent cutbacks, remains ten to one hundred times larger than is necessary to meet existing and projected national and international policy objectives, mainly deterrence of aggression.

To justify weapons programs during the Cold War, advocates spent enormous energy creating boogiemen and boogiegaps – for example, the missile gap, the bomber gap, and the Multiple Independently-Targeted Reentry Vehicle (MIRV) gap – none of which actually existed. Vast resources were spent as well, studying the effects of nuclear detonations and carrying out a number of ethically questionable experiments on human subjects that have only recently come to light. Astoundingly, all of the thousands of defense experts working for forty

FIGURE 20.1

The Chapleau, Canada, controlled fire of August 1985. The fire base is approximately 1 kilometer in diameter. The smoke plume has risen about twenty-thousand feet into the atmosphere, and the dense plume of smoke has blown downwind more than 100 kilometers from the site. (Photograph from R. Turco, 1995)

years appear to have missed one of the most important effects of all – global environmental and climatic change associated with smoke.

Everyone knows that fires produce smoke. An exceptional case is a well-controlled Bunsen burner, in which the flame is clean and blue. Figure 20.1, on the other hand, shows a large forest burn about one-half mile wide that is sending dense smoke high into the upper atmosphere. The fires at the oil wells in Kuwait during the Persian Gulf War reminded us that oil in particular produces copious black, sooty smoke, which consists of countless microscopic particles of carbon agglomerated into long chains and clusters. The effect of such smoke on sunlight in the atmosphere is predictable; in large enough quantities, it can turn day into night. What is not so obvious, but which Carl and others demonstrated, is that smoke can also cool the climate.

Taking the lead from Crutzen and Birks, the TTAPS team organized under Carl's leadership to investigate this important effect. TTAPS is the acronym formed by the initials of the surnames of the authors of the *Science* paper on Nuclear Winter – Turco, Toon, Ackerman, Pollack,

and Sagan [Turco et al. 1983]. I can recall the first meeting with Carl at
his fraternity house in Ithaca, New York, late in 1982. Brian Toon and
I had been summoned eastward by Carl, who was preparing for the
landmark satellite-linked conference, The World After Nuclear War:
The Conference on the Long-Term Biological Consequences of Nuclear War. Even though Carl had recently undergone serious surgery
and was obviously recovering from the trauma, his energy and enthusiasm for this important mission were overwhelming. By the end of
the visit, I found myself – relatively healthy and considerably younger
at the time – exhausted by the pace Carl had set. By our departure, Carl
had brilliantly set up the technical guidelines for the TTAPS study of
Nuclear Winter as well as for its subsequent exhaustive review.

The rest is pretty much history. Studies showed that a Nuclear Winter, though not certain, was possible. The most powerful predictive
global models available at the time were brought to bear, indicating
worldwide climatic disturbances. (Refer to [Malone et al. 1986]; see
also [Sagan and Turco 1990], Figures 16–21 and accompanying text.)
The primary cooling effect of the smoke occurs over land and may
reach 25° Centigrade or more in inland regions of continents. Along
coastlines, the cooling effect on land is strongly moderated by the
presence of warmer water nearby, particularly along western coastlines where prevailing winds carry warmer marine air onshore. The
predicted decreases in land temperatures – occurring in a matter of
days and weeks – represent enormous perturbations of the normal climate and could threaten human survival.

Carl introduced planetary science into the picture, noting, among
other things, that similar cooling had been detected on the surface of
Mars during large dust storms. The fact that Mars chilled when blanketed by dust strengthened the argument that Earth would respond
similarly when covered by smoke. The analogy between Mars' global
dust storms and Nuclear Winter provided a powerful image for the
desolation following a nuclear war (just as the ozone hole recently
offered a stark example of large-scale stratospheric ozone depletion).
Carl also started to ask hard questions about strategic nuclear policy
in the face of this potential global climate calamity.

In A Path Where No Man Thought: Nuclear Winter and the End of
the Arms Race [Sagan and Turco 1990], Carl and I attempted to develop
a simple, logical policy response to Nuclear Winter, keeping in mind
the uncertainties in global environmental effects [Sagan and Turco
1990, 1993; Turco and Sagan 1989a, 1989b]. In a Nuclear Winter the
impacts on climate and societal infrastructure are greatly magnified by
the fact that human survival depends on ample and stable food sup-
plies, while agriculture productivity is highly sensitive to global and
regional climatic change as well as to loss of high-technology human
support systems. A study of the likely outcome of Nuclear Winter on
the human population was published by the international Scientific

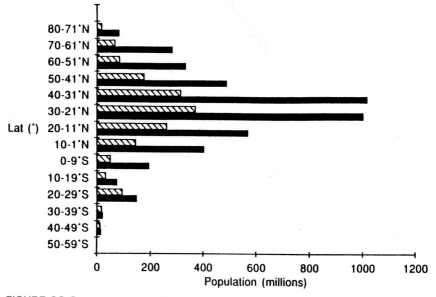

FIGURE 20.2

The human population currently living within specific latitude intervals, from north to south, is indicated by solid bars. Hatched bars show the estimated surviving populations after a full-scale nuclear war followed by Nuclear Winter impacts on agriculture, assuming food stores are at median levels and food distribution to survivors is optimal. In all, less than a third of the initial population might be expected to survive. For a case in which food stocks are initially low and distribution systems are disrupted, more than 90% of the people on Earth are predicted to perish. [Harwell and Hutchinson 1985, Figure 7.4, p. 480]

Committee on Problems of the Environment (SCOPE) [Harwell and Hutchinson 1985]. The SCOPE study suggested that most of the human population might perish after a nuclear war (refer to Figure 20.2) — a far cry from earlier, more optimistic extreme estimates that perhaps only a few hundred million people might die, which was deemed to be manageable. Obviously, the stakes in a nuclear war had changed qualitatively from manageable to unacceptable. Yet many policy experts did not see it that way. At a high-level meeting that Carl and I attended in Washington, D.C., soon after Nuclear Winter was first announced, one well-known nuclear strategist stated that "if you believe the threat of the end of the world will change thinking in Washington or Moscow, you have never spent any time in either of those places." In fact, there is some evidence to suggest that the putative threat of the demise of human civilization as we know it had indeed deeply affected individuals of conscience in high places all around the world.

The threat of mass starvation following a nuclear war remains today a possibility because of the large numbers of nuclear weapons remaining in the national arsenals. By our estimates, even a relatively small number of weapons could result in serious climatic effects, depending on how the weapons were used. For example, strategically valuable and vulnerable oil facilities, together with other major industrial assets, could alone generate enough smoke to disrupt worldwide agricultural output significantly.

Accordingly, to minimize the potential threat of Nuclear Winter – should the worst ever happen – Carl and I developed a deterrence concept of minimum sufficiency. Minimum sufficiency relies on a small, robust force of weapons amounting to roughly one hundred warheads on each side. Although such an arsenal is still equivalent to more destructive power than all of the conventional weaponry ever manufactured, it is far smaller than recently proposed drawn-down nuclear inventories of thousands of weapons. The advantages of very small nuclear forces are even more clear-cut in today's post-Soviet world, in which strategic global warfare is no longer a danger. Hence, the continuing drumbeat of rhetoric calling for more nuclear weapons, more weapons research, and more nuclear testing is mind-numbing.

To achieve small nuclear arsenals, Carl and I sought to lay out a road map for the weapons wonks. A sketch of our plan to phase out weapons and rebalance the superpower arsenals is illustrated in Figure 20.3. To be sure, we were cocky and overconfident in expecting military experts to take seriously such a bold and straightforward plan. And none of them, as far as I know, ever did. In the late 1980s, we were proposing a steady reduction in the numbers of strategic warheads to thousands by 1995, on the way to hundreds early in the next century. At the time, such a plan was considered outrageous! Today we are approaching the 1000s; but will we move toward lower numbers? Carl and others feel we must.

Complicating the picture was Ronald Reagan's Strategic Defense Initiative (SDI). This misbegotten project was always a red herring in discussions of strategic weapons. In the 1980s, the United States was wasting 5 billion dollars a year to realize a simple-minded dream of an impenetrable defensive shield in space, with the ulterior motive of spending the Soviets into bankruptcy. We nearly spent ourselves into the poor house instead. Carl and I (and many others) suggested that this porky boondoggle should be rendered down to size, a position the Pentagon itself had adopted. Yet there has been a dogged persistence to the concept of a foolproof defense. SDI's most recent incarnation is a space-based system of missile-borne nuclear explosives to defend the Earth from errant comets and asteroids. This flimsy pretext has been used to argue for more nuclear weapons development, testing, and stockpiling! The path that Carl pioneered – reduction of SDI to a modest-sized advanced technology program – makes a lot more sense.

The nuclear weapons testing program has proven to be an effective way of keeping nuclear scientists and technicians employed over four decades – first mucking up the environment with radioactivity

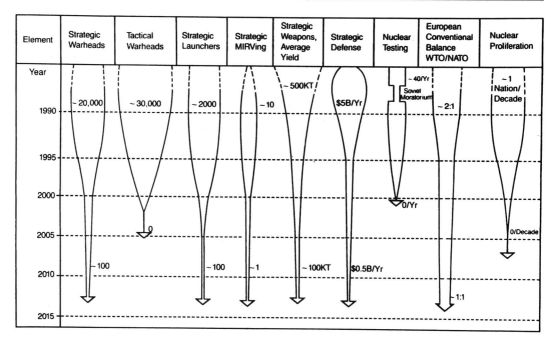

FIGURE 20.3

A road map for nuclear weapon reductions. The time lines for reductions in strategic and tactical nuclear weapons and for phasing out nuclear tests are given for the period from 1990 through 2015. Progress has already been made along these lines since 1990. [Sagan and Turco 1990, Figure 7, p. 293]

and more recently cleaning up the mess (which may take four more decades). In Carl's mind, the path here always led clearly to zero testing. And that is where we seem to be headed. Indeed, the last obstacle to reaching our destination along all of these various pathways is nuclear proliferation. Here, all parties agree, enlightened leadership will be the key to stem the spread of the technology of mass destruction.

In summary, it is clear to me that the strength of Carl's will, his broad scientific training and genuine creativity, and his deep concern for the welfare of the human species (and everything else on this planet, and indeed other planets yet to be discovered) have helped in a substantial way to draw civilization back from the brink of nuclear self-destruction.

BIBLIOGRAPHY

Crutzen, P., Birks, J. 1982. Twilight at noon: The atmosphere after a nuclear war. *Ambio* 11:114–125.

Harwell, M. A., Hutchinson, T. C. 1985. *Environmental Consequences of Nuclear War, Volume II. Ecological and Agricultural Effects*, SCOPE-28. Chichester: Wiley.

Malone, R. C., Auer, L. H., Glatzmaier, G. A., Wood, M. C., Toon, O. B. 1986. Nuclear winter: Three-dimensional simulations including interactive transport, scavenging and solar heating of smoke. *J. Geophys. Res.* 91:1039–1053.

Sagan, C., Turco, R. P. 1990. *A Path Where No Man Thought: Nuclear Winter and the End of the Arms Race*. New York: Random House.

Sagan, C., Turco, R. P. 1993. Nuclear winter in the post-cold war era. *J. Peace Res.* **30**:369–373.

Turco, R. P., Sagan, C. 1989a. Policy implications of nuclear winter. *Ambio* **18**:372–376.

Turco, R. P., Sagan, C. 1989b. Strategy and policy in a nuclear-armed world: Implications of nuclear winter. Report, Laboratory for Planetary Studies (Carl's center), Cornell University.

Turco, R. P., Toon, O. B., Ackerman, T. P., Pollack, J. B., Sagan, C. 1983. Nuclear winter: Global consequences of multiple nuclear explosions. *Science* **222**:1283–1292.

Turco, R. P., Toon, O. B., Ackerman, T. P., Pollack, J. B., Sagan, C. 1984. The climatic effects of nuclear war. *Sci. Amer.* **251**:33–43.

Turco, R. P., Toon, O. B., Ackerman, T. P., Pollack, J. B., Sagan, C. 1990. Climate and smoke: An appraisal of nuclear winter. *Science* **247**:166–176.

Turco, R. P., Toon, O. B., Ackerman, T. P., Pollack, J. B., Sagan, C. 1991. Nuclear winter: Physics and physical mechanisms. *Ann. Rev. Earth Planet. Sci.* **19**:383–422.

21

Public Understanding of Global Climate Change

JAMES HANSEN

NASA, Goddard Institute for
Space Studies, New York

In the middle of the 1960s, when I was a 24-year-old student at Kyoto University, Japan, on leave from the University of Iowa and feeling anxious to define some calculations that could serve as the basis for a Ph.D. dissertation because I already had applied for a postdoctoral position at NASA, I wrote a letter to Carl Sagan, a young Assistant Professor at Harvard University.

I had a notion that I would like to dispute Sagan's theory that Venus was hot because it had a thick greenhouse atmosphere. My idea was that Venus could be kept warm by a dusty atmosphere, which trapped internal planetary heat. He responded with a detailed explanation as to why he thought that was not likely. I made heavy use of many Sagan and Pollack Venus papers, but I persisted in developing my dust model and got my Ph.D. the next year, whereupon I ended up at a Kitt Peak conference on the atmosphere of Venus, at which Carl Sagan was a key participant. I was too shy to approach him, but I did talk with his student Jim Pollack, and we ended up at lunch with Carl, who asked me if I really believed that it was dust that kept Venus so hot. I said I thought it more likely to be the greenhouse effect, but someone had beat me to that idea, and I thought the dust model a plausible alternative. Of course, when the U.S. Pioneer spacecraft arrived at Venus they proved conclusively that Carl was right.

A year after the Kitt Peak meeting, Jim Pollack and I were sailing on a small lake in New Hampshire during a Gordon Conference – more precisely, we were floating on a becalmed Sunfish. I remember our discussion well. One topic was Carl Sagan's propensity to communicate science with the public via the media. We noted that some scientists seemed to consider this to be an aberration, we thought we perceived a bit of prejudice against Carl because of that, and, in our youthful idealism, we decried the injustice. We agreed on the importance of communicating science to the public and that it was a pretty tough job, no matter how easy and natural Carl may make it seem to be. But, in retrospect, I now realize that we didn't have the foggiest notion of just how complex and difficult such communication can be.

In the past few years I have gained a little perspective on this difficulty. As a lead-in to a climate discussion, let me mention a five-minute testimony on the greenhouse effect that I gave to a committee of the U.S. Senate in 1988. I made three assertions. First, that the world was getting warmer on decadal time scales, which I said could be stated with 99% confidence. Second, that, with a high degree of confidence, I believed there was a causal relationship of the warming with an increased greenhouse effect. And third, that in our climate model there was a tendency for an increase in the frequency and severity of heat waves and droughts with global warming.

This testimony received attention because of the heat and drought in the U.S. that summer. But in response to a question from one of the Senators I stated that no specific drought could be blamed on an increasing greenhouse effect; it only altered the probabilities. In a later testimony to Senator Gore's committee, I expanded this conclusion by noting that the intensity of both extremes of the hydrologic cycle, droughts and forest fires on the one hand, and heavy rains and floods on the other hand, would increase with global warming.

Leaving aside whether other scientists agreed with it, this seemed like a simple message. But it soon became obvious that the media and the public misunderstood it. The television program *Jeopardy* stated that I said the drought was caused by the greenhouse effect. The media's power must be pervasive. Just last week I read a preprint by Hans von Storch of the Max-Planck-Institut, one of the better meteorological laboratories in the world, in which he states that during the 1988 Senate hearings "James Hansen declared the drought with '99 percent certainty' to be related to the anthropogenic climate change." I suppose his ultimate source must have been *Jeopardy*; it certainly wasn't my testimony or my publications.

During my testimony I specifically illustrated that present global warming of half a degree Celsius is less than natural variability of regional temperature. But after seeing the media interpretation, I realized that lots of people would misunderstand it the next time the temperature in a given season was colder than normal.

So I made up this set of colored dice. The die with two red, two white, and two blue sides represents the period from 1951 to 1980. Red is for a season warmer than normal, defined by the temperature range of the ten warmest seasons of that thirty-year period. Blue is for cold seasons, and white is for near-average temperatures. My claim was that, with those temperature ranges fixed, the increasing greenhouse effect would shift the odds such that averaged over the decade of the 1990s, the chances of having a warm season would be increased to four chances out of six instead of two out of six. And I thought that may be a loading of the dice that would be sufficient for the man-in-the-street to notice that climate was beginning to change.

I used these dice on television programs a couple of times, but I'm not good at that stuff and I don't like it at all. I concluded that I could explain things a lot better by writing an article for a popular audience, so one Monday morning I called the editor of *PARADE* magazine and offered my services. He was very nice about it but politely declined and told me that he already had arranged for an article on the greenhouse effect by Carl Sagan. That gave me the great idea of sending Carl a set of these dice and suggesting that he use them to help explain the impact of greenhouse warming on a noisy climate. I'm not sure how much he has used them yet, but the Symposium for Carl reminded me about the dice, and I realized that we are already halfway through the 1990s. Therefore I thought it would be interesting to check how the frequency of warm seasons has changed, if at all, to see whether I gave Carl a bum steer.

So I got the temperature data for all the [MCDW, Monthly Climatic Data of the World] meteorological stations at middle latitudes in the Northern Hemisphere. Plate XV shows that at the beginning of the 1990s the occurrence of warm seasons briefly reached the level predicted for the decade average – 1990 was the warmest year of the century. After the eruption of Pinatubo in 1991, the largest volcano of the century, there was a cooling back to levels comparable to the 1950s. There is evidence in the most recent years for a rebound from Pinatubo cooling.

Plate XVI shows the percentage of warm stations for the entire globe, with each latitude zone weighted by its area. The fluctuations are not as violent, and it appears that even with Pinatubo the frequency of warm seasons has reached 50% – that is, three sides of the die are red.

I'm confident that by the end of the 1990s we will see that the average for the decade is on target. We have shown, for example, in a recent *Research and Exploration* article [Hansen et al. 1993] that there should be substantial warming in the last half of the 1990s. The reason is that anthropogenic greenhouse gases are already pushing the climate system hard enough to compete effectively against unforced climate variability. I'm confident that the record global surface air temperature level of 1990 will be exceeded at least once in the second half of the 1990s, probably more than once. I'm not sure whether my confidence will convince Carl to make more use of my loaded dice.

What is our confidence based on? The newspapers say that the predictions come from climate models – calculations on a computer. But that is naive and very misleading. Actually, expectations of climate change are based on understanding of the Earth's climate system derived from analysis of observational data, with the help of climate models.

A climate model is a tool that lets us experiment with a facsimile of the climate system, helping us to think about and analyze climate, in ways that we could not, or would not want to, experiment with the real world. Climate modeling is complementary to basic theory, laboratory experiments, and global observations. Each of these tools has severe limitations, but together, especially in iterative combinations, they allow our understanding to advance. Although models are very imperfect, they structure the discussions and help to define needed observations, experiments, and theoretical work. Perhaps it is useful to illustrate this with a contemporary example, the natural climate experiment provided by the volcano Pinatubo, which is just beginning to be analyzed.

The Mount Pinatubo eruption in 1991 injected about 20 megatons of SO_2 into the Earth's stratosphere, where it was dispersed by winds and formed into a global layer of fine sulfuric acid droplets or aerosols, which scatter sunlight back to space and also absorb the Earth's heat radiation, thus cooling the Earth's surface and warming the stratosphere. The reflection of sunlight by Pinatubo aerosols is the largest global climate perturbation this century, and it provides a valuable opportunity to test climate models and improve our understanding of the sensitivity of climate to such a global radiative forcing.

Plate XVII shows temperatures in the stratosphere, troposphere, and at Earth's surface for the past few years. The climate model calculations were made and published shortly after the Pinatubo eruption, under the assumption that the aerosols would be of the same size as was measured after the earlier volcano, El Chichon, and that the amount of aerosols was about twice that for El Chichon.

The observations show that the stratosphere did warm up quickly, but it didn't cool off as fast as in the model. We now have observations that show the Pinatubo aerosols continued to grow in size for more than a year after the eruption, unlike the assumed El Chichon aerosols, and that is probably the reason the stratosphere remained heated for a longer period. Also the stratospheric cooling evident before and after the eruption is clearly due to ozone depletion, which was not included in this simulation. We show in our *Research and Exploration* paper [Hansen et al. 1993] that the main effect of ozone depletion on temperature over the past 15 years has been a stratospheric cooling that is now about 1°C.

Because the Pinatubo aerosols were larger than we had assumed, we may have overestimated the net radiative forcing of the troposphere and surface, perhaps by 10–20%. Nevertheless the model's tropospheric cooling seems to have been in the right ballpark. The tropospheric response, in one sense, is the most important, because most of the atmospheric mass is in the troposphere. On the other hand, we are especially interested in the surface, because that's where people live. The surface, on the average, did not cool as much as the model predicted; specifically, there was very little cooling in the Northern Hemisphere winters of 1991–2 and 1992–3.

That relative warmth in the winter was mainly over Asia, and it possibly was related to Pinatubo aerosols. Kodera and Yamazaki [1944] have presented evidence that the stratospheric heating by the volcanic aerosols would alter atmospheric dynamics so as to direct more warm oceanic air over Asia in winter. That mechanism probably could not have been simulated by the model used here, because this model version has only one and one-half layers in the stratosphere. Another possibility is that the winter wind patterns are just not sufficiently deterministic; chaotic fluctuations of wind patterns are largest during the winter.

There is potential for learning a great deal from the Pinatubo experiment, as we define the aerosol forcing more precisely and examine the climate response with a succession of different climate models. Probably the best chance for finding a cause and effect relationship between Pinatubo aerosols and observed climate change patterns will be in the summer seasons, because that is when the zonal winds and the atmospheric dynamical fluctuations are the weakest.

Plate XVIII shows the temperature anomalies during the last four Northern Hemisphere summers. 1991, in the upper left, was before the Pinatubo aerosols had a chance to form and spread; it was very warm relative to the 1951–80 average. 1992 was very cool over the major summer land masses, where maximum Pinatubo cooling was expected. In 1993 it was still quite cool over the continents, and it was wet with record flooding in the midwestern United States; we must ask and investigate whether the probability for this climate anomaly was not increased by the Pinatubo cooling. In 1994, as expected, the temperature returned most of the way to prevolcano warmth.

Pinatubo, as a natural climate experiment, will provide an example of how climate models help us to interpret real-world climate change quantitatively. It will help us iterate and improve upon our understanding of climate forcings and climate sensitivity to those forcings.

How can we get the public to understand better the nature of scientific research: what we know and don't know about climate change, how it is normal to challenge every theory or interpretation, and thus test and improve our understanding, whether it's the temperature on Venus, climate change on Earth, or whatever? Part of the difficulty is the teaching of science in our schools, which emphasizes facts with correct answers, rather than involving students in a true research process.

This is one of the reasons that we started this year an Institute on Climate and Planets in which we have brought in students and teachers from four New York City high schools and five junior and senior colleges to work with us. One goal is education: to show students and teachers how the research process works by involving them in it and working together to take it back to the classroom. We also expect to make significant research contributions: the problems chosen are all at the leading edge of current understanding. And we want to provide underrepresented minority youngsters opportunities to develop their potential to contribute in science and gain access to career paths in research: all of the students are underrepresented minorities.

One of our four projects is called Pinatubo. My last two figures were produced by the students in the Pinatubo project, from Andrew Jackson High School, Bronx Science, City College, and York College. The first task we assigned them was to test the ability of our newest climate model to simulate the average climate of the real world. Plate XIX shows the difference between the model's winter surface air temperature and climatology. We were disappointed to find that this new model had an error as large as 10°C over Canada. You can see from the wind anomalies, overlain as arrows on the temperature map, that the warm air is related to errors in the tropospheric wind pattern. In the future the students will be checking how different alterations to the model's physics might improve the simulation. Of course there will always be model imperfections, so the model is also being used for Pinatubo experiments that will be repeated in the second and third years of this program with improved models. In this way we hope to identify which conclusions are relatively independent of model imperfections.

The second task we assigned the Pinatubo group was to compare the observed variability of climate with the unforced variability in the model – that is, the chaotic fluctuations of climate that occur from year to year simply as an unpredictable sloshing around of the fluids. Plate XX was produced by Andrew Jackson students. The top part is the standard deviation of observed surface air temperatures of the past fifteen years; these observations must include both unforced variability of the system – that is, noise or chaos – as well as climate variations due to deterministic forcings such as Pinatubo aerosols. The lower figure is the unforced variations, or chaos, in the model. We have evidence that the model's unforced variability is fairly realistic in the winter, and, as you can see, the unforced variability represents a large fraction of the total observed climate variability of the past fifteen years. This implies a severe limitation on the predictability of regional climate fluctuations. But it does not mean that we cannot find substantial forced climate changes after a major perturbation such as Pinatubo. Also, although the expected signal from increasing greenhouse gases presently is less than unforced regional climate variability, the greenhouse signal will grow steadily with time and eventually exceed this regional noise level.

That brings me to a question asked by the public: When do we expect humanmade global warming to be obvious unambiguously? I said earlier that I was confident that we would see the record 1990 global temperature level exceeded this decade, probably more than once. That will have significance, because, as Bassett and Lin showed in the journal *Climatic Change* [Bassett and Lin 1993], the recent cooling pushed global temperature far enough away from the 1990 record that it would be very unlikely for chance fluctuations to lead to a new record. In other words, a new record will represent evidence of a dominant deterministic climate forcing.

I would guess that such a new record will renew the heat of the climate debate, but it certainly will not silence scientific greenhouse critics, and indeed it should not. These critics are not a barrier to the advancement of scientific understanding of climate change. The scientific method invites continual criticism and reassessments of understanding. It thrives on this. That is how it advances. The main barrier to scientific understanding is the absence of adequate monitoring of climate change, especially of the forcing and feedback mechanisms that cause the changes. Uncertainties about climate change will exist for many decades, so we need to get in place the measurement systems that will allow us to understand the changes that do occur and thus be in a position to help evaluate the costs and benefits of relevant public actions and inactions.

But perhaps the most difficult matter will be attainment of a good public understanding of what we know and don't know. It will always be possible to find experts on all sides of any scientific issue, so it will be necessary for the public and their leaders to understand how the scientific process works, to weigh the evidence, and to choose among alternative actions. That's one reason that science education in our schools, for all of the students, is so important. And finally, it underlines the need for the rare person like Carl Sagan, who can help educate the public on such issues. We had better hope that he continues working for a good number of decades.

BIBLIOGRAPHY

Bassett, G.W., Lin, Z. 1993. Breaking global temperature records after Mt. Pinatubo. *Climatic Change* **23**:179–184.

Hansen, J., Lacis, A., Ruedy, R., Sato, M., Wilson, H. 1993. How sensitive is the world's climate? *Natl. Geograph. Res. Explor.* **9**:142–158.

Kodera, K., Yamazaki, K. 1944. A possible influence of recent polar stratospheric coolings on the troposphere in the northern hemispheric winter. *Geophys. Res. Letters* **21**:809–812.

Monthly Climatic Data of the World. 1995. Asheville, NC: NOAA National Climate Data Center.

22
Science and Religion

JOAN B. CAMPBELL

National Council of the Churches
of Christ

Science and religion. Each one claims enormous human energy, power, and endless intellectual attention. Someone had a demonic sense of humor to allow me a mere few pages to discuss such an important subject. Nevertheless, the brevity is a blessing, for even in a longer paper no one could adequately address this topic. I have observed that virtually every contributor to this volume, including Carl and those who asked questions of him after his public lecture of the birthday symposium, have noted the existence of the world of religion.

Let me begin by trying to draw in the parameters of this large subject. The final section of this book has been designated "Science, Environment, and Public Policy," and I would like to focus on our communities – the community of science and the community of faith. Someone quipped that the evening party preceding Carl's birthday symposium was a collegium of graduate students several decades after; a class reunion; a warm, friendly, and vigorous community. In fact, the three of us in the small religious caucus (Dean James Morton of St. John the Divine in New York City, my colleague Dr. Albert Pennybacker, and myself) commented to one another that we were actually more comfortable in this gathering than we would be at a gathering of business leaders in the Chamber of Commerce. "Or," as the Dean said in a reflective moment, "perhaps in a gathering of church hierarchs." The communities of science and religion do not always or even often merit a favorable comparison. Throughout history our two communities have been seen as antagonists. Yet it is the possibility of a shared community that brings us together to address our present circumstances and our unfolding future.

This is not simply a matter of cutting the subject into a manageable bite. It is that, but it is also a recognition that old debates are essentially exhausted. To a large extent, old antipathies have been laid to rest, at least in the progressive religious community. Perhaps this has occurred for good reasons: a maturing of religious thought and a certain integrity, or even humility, in science's understanding of itself (which may be its sign of maturing). Antipathies also retreat in the context of realism about the major shift that has occurred. We now live in an age of science. At one time science struggled in an age of religion. Ours was the ascendancy, and we did not handle it very well. We hope you will be more successful in this time of the ascendancy of science, and we will encourage that. (Dare we say we will pray for that, or was that what got us into trouble in our time?)

But beyond antipathy – out of the ancient, continuing place that religion holds and the ascendancy of science now comfortable with itself – there is the possibility of sharing our engagement with our common life and seeking a better way.

Let me tell you what I believe is an important story. It is a story of a partnership, and it began with what Carl has described to me, at least apocryphally, as a direct, empirically verifiable, peer-reviewed experience of divine revelation! This somewhat secret story is about Dean James Morton's efforts to get the Episcopal Church to be more concerned about the environment. The Dean thought he had been successful when his church agreed to discuss stewardship nine years ago. Unfortunately, on the way to the convention they turned it into a discussion about funding, not about the environment. Determined to find a new strategy to capture church people's attention, he decided to issue a challenge to them from the scientists. Rather than choosing an encyclical, a movie, book, or television series, a letter was composed. It was titled "An Open Letter to the Religious Community," and history will mark it as a key, catalytic event that led to the permanent, irreversible integration of global environmental issues into mainstream American religious thought and life.

Few, perhaps no one but Carl, could have found the perfect refinement of tone that would communicate such authority, authenticity, and activism. Few could have identified and persuaded in just a few weeks thirty-two scientific colleagues of enormous stature to add their signatures. In the letter, the scientists said that humankind was close to committing (many would argue we are already committing) what in religious language would be called crimes against creation.

As much as you can encapsulate a breakthrough in a few words, the letter went on to say, "As scientists, many of us have had profound experiences of awe and reverence before the universe. We understand that what is regarded as sacred is more likely to be treated with care and respect. Our planetary home should be so regarded. Efforts to safeguard and cherish the environment need to be infused with a vision of the sacred." Lest you lose track here, these are the scientists speaking to the religious leaders.

That combination of urgency and deep recognition – across the distance and history that we share – of the dimension of the global crisis made the message strong and convincing. In a few more weeks, several hundred religious leaders cosigned a letter expressing immediate willingness to undertake earnest dialogue. The exchange of letters was formally announced at a meeting convened by President Gorbachev in January 1990 at the height of perestroika. Such was the atmosphere of openness, hope, and possibility in which the letter was presented. Though I was not there, it is rumored that Gorbachev, and certainly Scheverdnadze, joined in a chant invited by the Hindu priest in which all repeated the sacred word "om." The story is told that later that night, and it was a Friday, a minyan of Jews celebrated Shabbat in a small room together for the first time within the walls of the Kremlin.

This letter led to two years of activity under a process known as the Joint Appeal by Religion and Science for the Environment, now called the Religious Partnership. Those discussions had many memorable moments. Once we were seeking to prepare another letter to present in the U.S. Capitol. The document was to be signed by Carl and the scientific community but also by representatives of the Southern Baptist Convention and others who are religiously conservative, not often a part of such ecumenical ventures. Some expressed concern about words that referred to global warming at a rate unprecedented in tens of millennia. "Some of us don't think we've been here that long," they said. Carl helped draft new words of consummate common sense: "We do not have to agree on how the natural world was made to be willing to work together to preserve it."

What does it take sometimes to set in motion significant movements of thought and action? One night, some years ago, one man, our friend Carl Sagan, decided that a letter needed to be written.

I believe that in our communities, in all our humanity, we are finding ways to address together the human predicament, the threat to this wondrous environment, and the diminishing of life as we know it. Good science and good religion are making book! Bad science and bad religion have found it relatively easy to make book. For both it is usually rooted in self-service, dangerously so. Science can be socially naive, without ethical inquiry and restraint. Religion runs the risk of being unreflective, without integrity, or intent on lining its own coffers. It can become a theocratic movement that is neither harmless nor innocent. Imagine absolutist religion married to the power of our modern technology. It is a prospective but very possible nightmare. Our common life would become seriously threatened, and that matters. It is wrong and we all know it. It is up to the scientific community to say what is good science. I have no qualifications to speak to that. I can, however, speak about good religion, and it is good religion that needs to put bad religion in its place. Here I want to draw on the best of theological reflection, a field of disciplined, mature inquiry and thought about which many in the field of science cannot be expected to be broadly informed.

Good religion talks about ultimate, or primary questions. When asked to define religion, Langdon Gilkey, a theological witness at the Arkansas trial on creation, suggests that religion holds to a certain view of the nature of reality; ultimate reality; reality as a whole. Further, religion centers its attention on the relationship of that ultimate reality to the deepest problems of men and women and even of nations: sin or alienation, finally from life itself, injustice, the abuse of life; death and rebirth. That is, religion addresses the question of meaning in existence and it answers in terms of symbols, myths, teachings, scriptures, doctrine, and dogmas. When it is good religion, it answers in ways that resolve the deepest of human problems and build up the community of life [Gilkey 1985, pp. 99–100]. Good religion does not build walls that divide.

Science is also a way of knowing reality. Perhaps it is our most reliable and fruitful way. Science is a wondrous power, engaging in proximate and largely immediate reality. It abstracts and objectifies and thereby equips us with the ability to understand and function in the midst of this setting for life that we have been given. Science helps answer the how maybe more than the why questions. Gilkey illustrates: "When it rains, we turn to the meteorologist to find out what caused the rain, and how it came into being and how it passes through. But when the bride asks, 'Why is it raining on my wedding day?' that is a religious question" [Gilkey 1985, p. 122].

Further, religion encourages a way of life steeped in renewal, redemption, and rebirth, which in turn provides the freedom to care, to risk, and to commit. That is, religion is found in lives that understand themselves related to God, the ultimate reality that religion affirms. In a way that parallels Carl's words in his paper, religion battles the "human conceit" and stands ready to dispute the philosophical claim that man is the measure of all things. Religion quotes the psalmist in the scriptures: "What is man that thou art mindful of him?" (Psalm 8).

Religion at its best leads to a sense of trust and a wondrous gratitude for the human capacities to share joy, to know love, and to live in relationship to the ultimate reality that it affirms. "Amazing grace," we sing, and in such dislodging of the human conceit, the human capacity is truly freed.

H. Richard Niebuhr spoke of this way of life

Revelation is not something miraculously esoteric; it occurs in various spheres of human experience. Faith is not an irrational leap into the absurd; it is confidence and loyalty, aspects of common human experience. Moral life can be represented by the human activities of 'makers,' 'citizens' and the 'answerers,' not only by the activities of moral philosophers. Ordinary human agents need no professional credentials to love and to care, and this is the way of life that matters [Gustafson 1994, p. 885].

For Carl, Annie, and I, and those who have joined in this partnership, this is the point of our coming together. This is the way of life that Carl spoke of where we take stones from one another's hands. I believe we are talking about the same vision. We come with different words, different disciplines, real disagreements, genuine respect, yet "the ground has been laid," as Martin Buber would say, "for real and genuine dialogue." I believe that the ecumenical religious community can find a shared sense of life with the scientific community. We can become partners.

Let me risk being teacher for a moment. When you read, "religious leaders say . . . ," be advised that the media is not very interested in moderate religious voices. Using your scientific skills, investigate such claims rather than settling for stereotypes and prejudices. There are some religious voices and communities that are different, that are inclusive and ecumenical. These are the people that want to stop the book burnings and fight the religious control of public schools; who defend religious liberty because without it, there is no liberty at all. These are the folk who, with an eye for a sad history, insist on the separation of church and state, who oppose prayer in public schools, religiously regimented study, and every other officially privileged status for religion. And they oppose it on deeply religious grounds. When hundreds of HIV-positive Haitian refugees came to our shores, it was this community that settled every single one, in spite of our government's claim that the refugees should be sent back because no one would settle them. It is this community that has been jailed at the South African Embassy, that presently risks the wrath of the religious right, that has ordained women and challenged patriarchy, and that awaits your scientific insights so that we might end our cruel homophobia.

What are some elements in our common ground? Already our experience with the Joint Appeal suggests several. Let me underscore three:

First, we share an emotion about the universe's beauty, mystery, and energy. We share a sense of wonder, respect, and affection for the world. Science has probed, assessed, and described with disciplined insight what the universe is in its unfolding. At no point is mature religion shattered or even compromised by such scientific insight. Our common ground is alluded to in Carl's novel, *Contact*, when the person of faith says to the scientist: "You have made the universe large enough for the God in whom I believe." Ours is a common awe before the majesty of creation.

Second, science offers facts for men and women of faith. Religion has not been formed by the disciplined, evidential grasp of facts that science provides. For instance, a scientist here at Cornell wrote to me in anticipation of what I might say. In a thoughtful piece he called into question the way in which religious claims about the afterlife ignore the reality of bodily death. He pointed out the total absence of any credible evidence that anything in the human body survives, It made me recall the rhyme:

> "I had a dog; his name was Rover;
> When he lived, he lived in clover;
> When he died, he died all over."

Facts have to chasten religious claims and insist on accountability. The question of the afterlife is a lecture all to itself and belongs to another occasion. The point is made, however. Religion and science have to deal with the facts that science with integrity provides.

Third, facts are critical. They are not morally and ethically neutral. Their accumulation, human knowledge, is not finally a bystander in life. Knowledge leads to power. This is also where religion and science meet. Once there is power, our question becomes one of ethics and morality: What shall we do with it? How shall we act? If religion is a realm of moral and ethical reflection, that reflection must also become a common ground for science and religion.

Out of these shared elements of our parallel life – emotional wonder, factual knowledge, and moral and ethical reflection – we can and must find a way to address together the condition of life, often the desperate conditions.

Let me be practical. You have facts that are the product of disciplined thought and the hard work of investigation. We have a constituency of committed people who care. Suppose through ecumenical religious access, we mounted a program where every congregation welcomed to its pulpit a person of science to speak the facts urgently and passionately. Would that make a difference? What would it mean for the morally committed to be confronted with the facts and address what is happening to our world and our universe? Religion is pulled toward a livable future for all just as much as it is driven by its tradition and its memory. Science is as well. A living partnership could be enriched and grow.

Does that sound too hopeful? Let's step aside and look at what has happened. Think of the image of planet Earth as a pale, blue dot or, in its more familiar form, that wondrous sphere filled with blues, grays, and purples. Ponder its uncanny resemblance to the ultrasound view of a mother's womb in the early stages of pregnancy. This image, a secular icon as it might be called in religious language, has helped form in our minds and hearts the reality of a world without walls; a world where barriers are broken and life breaks forth free and unfettered of the conceits of race, class, sexual preference, gender, and even nation-states.

Does it matter that the image of the world as womb hangs in a central place on the wall of the Vice-President's office? It is hard to know, impossible to prove, but something is happening. In less than a year, Arafat and Rabin exchanged the handclasp of peace. Yeltsin and Clinton stood together in the Rose Garden and talked of peace and cooperation. Mandela, thirty years a prisoner at the hands of the apartheid regime, took his place as the first black African President of South Africa, and 400 years of white domination fell, an event every bit as significant as the fall of the Berlin Wall. There is Ireland as well; and tomorrow Aristide will board a U.S. plane and return to power in Haiti. Old animosities are set aside, at least for the moment.

Who is to say what role the image of the pale blue dot played in these momentous events? Maybe none, but as for me and my house it makes a difference that science has given us the ability to see our tiny insignificant world as a "tableau rosa" on which peace might replace the rivers of blood.

BIBLIOGRAPHY

Gilkey, L. 1985. *Creationism on Trial.* Minneapolis: Winston Press.
Gustafson, J. M. 1994. Remembering H. Richard Niebuhr: Faithfulness. In *The Christian Century. Vol. 111.* Chicago: Christian Century Foundation.

23
Speech in Honor
of Carl Sagan

FRANK PRESS
Carnegie Institution of Washington

Thomas Huxley once said: "A man of science after the age of 60 does more harm than good." It may apply to some of us but Carl is one of the few exceptions! I know Carl from his public works and his work as a scholar in science – having participated with him on the Apollo project as working scientists and reading the scientific journals in which he publishes. We all know Carl as one of the great generalists of our time – an endowment that makes some narrow specialists wary. But with courage and verve Carl acts on the recognition (to quote Einstein) that "All religions, arts and sciences are branches of the same tree. All these aspirations are directed toward ennobling man's life, lifting it from the sphere of mere physical existence and leading the individual toward freedom." In his way Carl is one of our best exemplars of how one can break down the parochialism of expertise, yet be in the forefront of science and use that talent in a socially constructive manner.

Unfortunately, too many scientists tend to be impatient with concepts of the history of science, social consciousness, and even ethics. They care little about public outreach. This is in part because they had very little exposure in their training, in part because that's not where their bread and butter is, and in part because they see no gain. In French the word for popularization is vulgarization – which sums up the attitude of many scientists toward the popularization of science.

I would like to discuss briefly scientists and social conscience and conclude with some aspects of the sociology of science if there is time. I will do so, not by presenting a philosophical discourse but by giving examples of social consciousness done well, and done badly with destructive consequences. The main point I want to make is that science can only realize its full social potential for improving the human condition if it operates in a democratic society, if its practitioners receive a broader education in science than is often the case, one with underpinnings in history, culture, and ethics. And, finally, scientists must connect with society at large. To quote Carl: "In all of the uses of science it is insufficient to produce only a small, highly competent, well rewarded priesthood of professionals; some fundamental understanding of the findings and methods of science must be available on the broadest scale."

In the history of science an underlying theme has always been the aspiration to define a systematic procedure, a scientific method and ethic to uncover new knowledge, a process independent of individual temperament, and the cultural, social, and political pressure of the time. Of course this was never to be and cannot be. Science matters too deeply in military and economic affairs and in the standing of nations. And scientists, as they practice their profession, cannot be fully separated from their values. I need only remind you that Sakharov and Zel'dovich, Oppenheimer and Teller, and Heisenberg worked to develop nuclear weapons for governments with opposing ideologies that meant life or death for millions of people.

Some of you may have read about that celebratory party after the first successful nuclear test where Oppenheimer found a young group leader being sick in the bushes outside. He knew why. Science – pure, questing, removed – had indeed, known evil. And those who were there, from that group leader, to Oppenheimer, to some from Cornell who were also on that mesa knew that the ethic based on a clean separation of science from national goals was a fiction.

Yet that is not the whole story. After that experience scientists spoke out as experts and concerned citizens as never before. The FAS was created. There was a succession of nongovernmental studies of arms control and defense, spearheaded by scientists just as expert, and ultimately more influential than those inside the Defense and State Departments. The 1963 ban on atmospheric testing was sparked and promoted by the scientific community. Linus Pauling was awarded a second Nobel Prize for his efforts to stop atmospheric testing of nuclear weapons. Sakharov embarked on his courageous efforts in the Soviet Union at great personal sacrifice. He lived barely long enough to see the fruits of his labors.

That glimpse of the history of physicists and their developing social consciousness could be retold in the story of how the scientific foundation of the environmental movement evolved, in how the discoverers of recombinant DNA were the first to raise cautions and propose regulations of this new science until more was known, in the early warning on global climatic change, or the ozone hole.

There were difficulties faced by those in the vanguard of the developing social consciousness of American scientists. Scientists had to communicate their concerns to frequently hostile public officials. Scientists' tolerant attitudes toward each other often disappeared when it came to political differences. The public was often confused when scientists argued both sides of a public policy issue. Many scientists learned the often unrecoverable costs to their careers and to their scientific, and sometimes public, reputations of spending time in the corridors of power rather than in the laboratory, or of becoming concerned with public policy issues rather than in the number of scholarly papers they had published. And finally, scientists learned how slippery social issues became once they moved beyond stating fundamental facts derived from science to their policy implications.

I recall how physicist Mal Ruderman's home was picketed and how Murray Gell-Mann in Paris and Sid Drell in Corsica were shouted down while trying to deliver lectures in physics. Sid's social consciousness over the years was manifest in his involvement as an outside adviser and critic in arms control and defense issues and in other good works. Writing in the *Bulletin of the Atomic Scientists* he explained why he did what he did this way: "I think all men of conscience and intelligence face obligations associated with their knowledge and its potential effects on fellow citizens throughout the world. Some may choose to act solely through their scientific teachings and writings, others through their involvement with their governments, and still others through international organizations seeking to promote better human conditions. For myself, I have chosen a course which, *inter alia*, includes substantial efforts to affect in whatever way I can the policies of the United States through various scientific and technical advisory and working mechanisms... Since I live in a country in which I am privileged to have the opportunity to elect my government representatives, I have accepted the obligation to try to help the government function."

Nevertheless, I have heard it said by colleagues that science is a guide to asking the right questions, and these exclude social questions. And by right is meant questions that can be answered – by observation and experiment. Thus, science refrains from asking questions that it cannot ultimately answer. And it can't answer questions that go to moral and religious imperatives. Natural science can't ask what the universe was like before the Big Bang. Or why there is evil in the world. Or why nations and peoples fight each other.

Does it follow that a social consciousness based on scientific knowledge is inappropriate for scientists? Not so when an important social or political issue has a significant technical component. And the number of such issues is increasing rapidly in this world of growing complexity. To name a few: population control, the ozone hole and the banning of chlorofluorocarbons, health cost control, trade policy, security in the post-Soviet era, the control of carbon emissions, the regulation of fisheries, the reauthorization of the Endangered Species Act currently enmeshed in a political stalemate in Congress. Some 50% of the new legislation in Congress has an important scientific or technological component. I agree with Sir Crispin Tickell, former British

Ambassador to the United Nations (who is said to be responsible for the greening of Margaret Thatcher) when he wrote in the *New Scientist*: "Scientists should be much braver. I think this ethics argument – should they speak or shouldn't they – is a lot of crap. Scientists cannot promise certainty any more than economists can when they call for changes in taxes or interest rates. Uncertainty is part of the human condition. Caution in any case may in reality be recklessness. We must always look at the cost of doing nothing."

An historic event took place in Delhi, last year. Some fifty of the world's academies of sciences were convened by our National Academy of Sciences, the Royal Society, and the Swedish Academy of Sciences. This had never occurred before. They came together to see if they could agree to a statement on population addressed by the scientists of the world to the political leaders of the world. They did indeed issue a common declaration on population growth. It played a prominent role in the consensus that was achieved in Cairo in 1994 at the United Nations' conference on population. Simply stated the scientists of the world told their political leaders that science could not prevent the hunger, disease, environmental degradation, and dislocation that would ultimately follow from unlimited population growth. It outlined steps that had a scientific basis yet were consistent with humane and compassionate policies to reduce the birth rate. With this initial success I would hope that the world's academies, working together, would become a voice of reason as the nations wrestle with other world social issues.

With all of this I have a caution. It is exemplified by William Shockley using the credibility engendered by his great discovery and Nobel Prize to advance his social views on the racial inferiority of African-Americans. A current example is a distinguished mathematician, Igor Shafarevich, using the prestige of membership in the Russian and American Academies of Science to lend credibility to his antisemitic views and his entreaties for the ethnic purification of Russia. These are examples of parochial experts going beyond their special knowledge and taking prominent public positions on social policy. Social commentator and pollster Daniel Yankelovich had this to say about such interventions by ignorant, yet accomplished, specialists in an essay entitled *You Can Argue with Einstein*: "These examples help us realize how severely limited a learned person's experience may be. We have come to learn that experts, however impressive their credentials, often do not have an equal grasp of all modes of knowing." I wish he had chosen another title for his essay but I couldn't agree more with its contents. This is not to say that scientists cannot take up political or social causes remote from their expertise. But they should do so with no special claim to truth and virtue flowing from their scientific accomplishments. On the other hand, scientists with professional knowledge related to a public issue, whether it be nuclear war, environmental devastation, the dangers of a new technology, or whatever – operating in a democracy where contrary views can be heard – should speak out on these public issues and thereby enlighten the discussion. Linus Pauling did, Sakharov did, Jerome Wiesner, Sid Drell, Carl Sagan, Hans Bethe, Maxine Singer, and many contributors to this book continue to do so. What do all of these scientists have in common? The ones that I know have a sense of the history and culture of science, have broad scientific training, and a heightened social consciousness. By their intervention they elevate the level of public discussion to the credit of the scientific community and to the benefit of people everywhere. Again – let me repeat that scientists can be particularly destructive when they roam beyond the boundaries of special knowledge into areas where they are ignorant. This is especially dangerous when they represent the views of governments that brook no opposition. Remember the destruction of Soviet agriculture by Lysenko or the condemnation of Einstein's theory of relativity by both Nazis and Communists.

I would like to briefly take up two issues important in the social foundation of science that merit discussion on every campus. The first deals with the education and preparation of the next generation of scientists. The second relates to how a nation allocates its resources to science.

The system of training that most of our young scientists receive has evolved from the tough, competitive, peer review process that has emerged in the postwar years. It is true that this system has gained for the United States a world leadership position in science. But it has also produced too many young scientists who are overspecialized and illiterate of science as a culture with a history and with many dimensions. I believe it is time to reconsider a graduate school and postdoctoral process that turns out narrow specialists, ill prepared to deal with the economic transition, the restructuring of industries and universities, and the ethical issues that they will face as working scientists. It does not serve a young scientist well if his or her qualifications limit their job potential to an exceedingly narrow subdiscipline at a time when fields are changing rapidly, science budgets have changed from exponential growth to growth with the GDP. For many young scientists the professor as role model for a career may not be viable. Why can't the training of a particle physicist or astronomer be broadened so that he or she can also qualify as an engineer or applied physicist or data analyst so that there are more options for a rewarding and satisfying career in a rapidly changing society?

Too many newly minted particle physicists know little of the broad sweep of physics; too many geologists are uninformed of the problems society faces because humankind has become a more important agent of geological change than nature; too many young biologists have gotten into trouble because the code of ethics that evolved over history with the scientific method was never discussed in their training programs. It concerns me that scientists-in-training spend little time absorbing the broad cultural and historic context of science as it evolved between the time of Francis Bacon and Karl Popper. Why do we have so few Bethe's and Sagan's – first-class scientists who find time to be involved with issues of paramount importance to the nation like arms control and global climatic change?

Perhaps the situation is improving. A major review of the nature of the Ph.D. degree is underway at the National Research Council. I am particularly proud of a booklet called *On Being a Scientist*, published by the National Academy of Sciences during my term as president. It is addressed to graduate students and it takes up some of the historical and ethical issues that a literate and cultured scientist should be concerned with. More than 100,000 copies were distributed to graduate schools for individual reading and use in seminars. A new version is now nearing completion. The National Institutes of Health now require grant recipients to provide such training. One would have hoped that faculties would have recognized this need on their own. Nevertheless, this pressure is responsible for the growing number of courses in science history and ethics that are springing up in university curricula.

In discussing the allocation of resources for science I'll focus on two types of support, representing extremes in cost and administrative complexity. These are support for the individual researcher and support for science in the nation as a whole. Most of the grants that are made to individual scientists proceed through peer review. Peer review is hardly new. It dates back to the seventeenth century, when the gentlemen scientists of the Royal Society began to review articles submitted for its journal. So, the system has had a few hundred years to mature and to improve. In that light, we shouldn't be surprised that it works so well. The issue is not to abandon peer review. Rather, it is to make sure that we use it effectively.

First, individual research support. Some general principles apply. The awards are made to individuals, not organizations. Support is based on quality, not rank or affiliation. Support is on a cost-reimbursement basis. Finally, the research itself is monitored through the review process, through publications, and the use of research results by others.

Although individual research grants account for less than half of federal research support, they constitute the pillar of American science. And the efficacy of peer review – its efficacy in screening for excellence – is clearly critical to the strength of that pillar.

The concerns we have with peer review of individual research grants include both the new and the seemingly timeless. We worry that researchers are forced to become paper entrepreneurs – that the documentation for research proposals, the use of single-year grants, and the often multilayered review systems impose a time-consuming and costly burden. And for some scientists, the system may impose a seemingly impossible roadblock. What does that paper system do to the chances for a young investigator without a track record?

There is – rightly – a concern about undue conservatism in awarding grants. As the economist, Roger Noll, once put it: If one is in the business of buying science, how can one support a revolutionary idea? Indeed, many national review bodies have worried about the "traditionalism of peer reviewers" and of their "inability or unwillingness to recognize and recommend support for highly innovative, high risk proposals."

Noll tells a story that I think frames the issue nicely. He asks you to imagine yourself as a staff member at the National Science Foundation. You have just received a fat proposal from a twenty-nine-year-old physicist. In it, he asserts that "all the work in biology up to now is rotten. I have a totally different way of looking at it, which starts in physics. My physics background is sufficient for me to revolutionize biology." Suppose you also know that this physicist's idea of fun is to drive his car into the desert until he gets stuck and then pitch his tent. As Noll put it, "The probability of that proposal getting support is zero." But, of course, the physicist was Max Delbruck. And he did have a major impact on biology, an impact for which he won a Nobel Prize.

There are other concerns with peer review of individual proposals. For example, do potential grantees write their proposals to fit the review system or the committee structure rather than to lay out their best ideas? Are researchers forced to split coherent long-term programs into smaller, unconnected projects? And what is the effect of that?

The system of peer review may force scientists to think in traditional and established ways. As the impacts of science rapidly begin to cross disciplinary lines – as physics affects biology, as computer science affects physics, as genetic advances affect agricultural science – innovative researchers continually seek ways to apply and use the transfer of knowledge. But how does a trail-blazing scientist satisfy disciplinary-oriented review panels – and, indeed, disciplinary-oriented journal editors? I am delighted that the nation's largest granting agency, whose new Director knows that I'm talking about the National Institutes of Health, is now reviewing their process of peer review with these kinds of concerns in mind.

With regard to overall science policies I might remind you of the three periods of the history of Science and Technology Policy in the United States. The first corresponds to the period before World War II when the federal government was a minor supporter of science and technology, and most of the nation's research was carried out in the universities supported by philanthropy and a few industrial laboratories. This is the period where Europe was the center of scientific leadership, the United States led in technology and was the dominant industrial power. One might say that the United States was the Japan of that period.

The second period is the one that began with the end of the war and is ending now. It is appropriate to call this the Vannevar Bush Era because his influential report of 1945, *Science, The Endless Frontier*, set the stage. This era is, of course, the one I and a few of you grew up in, the one in which world science was led by the United States. It is characterized by the federal government's predominant role in the support of fundamental science and engineering in universities and federal laboratories. This is also the period the United States became a world leader in science and military technology but lost its primacy as an innovator in product design and manufacturing in many important sectors.

We have entered a new era, which I call the post-Vannevar Bush Era until someone thinks of a better name. I believe that in this era the United States, Japan, and Western Europe will all strive to be strong in science, technology, and excellence in design and manufacturing and global salesmanship. President Clinton's major statement on his science and technology policy, issued last month, deals with this new era. To the credit of the administration it recognizes the intrinsic value of basic research and does not require a test for relevance or contribution to some national purpose. It also recognizes the need for efficient mechanisms to exploit new knowledge to the economic and social benefit of the nation.

When I was in the government the then director of the Office of Management and Budget asked me (in front of the President): How do you scientists know when you have enough money. The question is even more appropriate today and I believe we have an answer that fits the times. It comes from the last report that was issued by the National Academy of Sciences during my term. The Committee on Science, Engineering and Public Policy of the Academy undertook a broad reexamination of the federal rationale for investing in science and technology. Some of the recommendations in the report were included in President Clinton's policy statement that I mentioned earlier. The report argues that the sole comparative advantage of the United States in the years ahead will be its scientific strength. Yet, it was a statement that recognized that exponential growth in the numbers of scientists at work and the financial resources to support them could not be sustained. It raises the question of how many scientists does a nation like the United States really need. It sets out performance goals that for the first time provide policy makers with a yardstick to gauge how much to invest in science. This is the gist of the recommendations:

- "The first goal is that the United States should be among the world leaders in all major areas of science. Achieving this goal would allow this nation quickly to apply and extend advances in science wherever they occur."

- "The second goal is that the United States should maintain clear leadership in some major areas of science. The decision to select a field for leadership would be based on national objectives and other criteria external to the field of research."

This may sound chauvinistic, but the world looks to the United States, as the only remaining superpower, to act in that role! Leadership in select fields is essential for that role. My choice for leadership for these years would be astronomy for the intellectual content and pace of discovery, biology for obvious reasons, material sciences including related fields in engineering, condensed matter physics, and chemistry for the economic impact, and the earth sciences because of their connection to global environment and resource issues.

Each country has to find its own route to improve the standard of living, the security, the health of its people and to see that the cultural and intellectual life of the nation thrives. For the United States with its many problems the path is made easier because of the great research university system that it has built and the world leadership position in science that has resulted. Carl Sagan said it this way, writing in the *Washington Post Book Review*, January 9, 1994

> [Science] makes the national economy and the global civilization run. Other nations well understand this. This is why so many graduate students in science and engineering at American universities – still the best in the world – are from other countries. Science is the golden road out of poverty and backwardness for emerging nations. The corollary, one that the United States sometimes fails to grasp, is that abandoning science is the road back into poverty and backwardness.

EPILOGUE

24
Carl Sagan at Sixty*

FRANK H. T. RHODES

President Emeritus, Cornell University

Carl and Annie; members of the family, whom we have just honored and are so proud to see here this evening; friends. I want to bring Carl greetings from all the members of the Cornell community who owe so much to his leadership over so many years. It is remarkable that it takes two days and twenty-five speakers to cover the mere outlines of Carl's work. Even more remarkable are two things that seem to embody the character of the symposium and that are so evident here this evening. One is the astonishing breadth of fields – astronomy, science education, public policy, and the relation between ethics and science – to which Carl has made a contribution and which have engaged this large number of people from varied backgrounds for two days. The second is a quality that doesn't always mark professional symposia: a real sense of love. There has been a sense not just of respect and admiration, but of love in this room tonight. That is a remarkable thing.

I want to say happy birthday and thank you to Carl as an exemplary member of the Cornell community. It will not have escaped those of you who spend your days in the academic vineyard that the scholarly profession, the academic life, the professorial calling have had something of a bad press in recent years. On Friday I had the privilege of speaking to our own assembly of faculty members about the outlook. We talked together about the academic profession and the need to state explicitly what our expectations are as professorial colleagues. I want to salute Carl Sagan tonight as the embodiment of everything that is best in the academic life and to explain why we are so immensely proud to be able to call him colleague here at Cornell.

We require three things from any faculty member: scholarship, teaching, and service. And, however you analyze the equation, Carl's

---

*Banquet address delivered on October 14, 1994, at Cornell University, celebrating Carl Sagan's sixtieth birthday.

performance in each has been stellar. As a scientist, he's a man of extraordinary breadth, partly no doubt because his own heritage of astronomy, biology, physics, and chemistry come together in a wonderful convergence. He has carried even those of us who are not astronomers all the way from the origin of life on this beautiful planet to the puzzling problems of the seasons on Mars and windblown dust that goes with them, to the surface temperature of Venus and the greenhouse effect that produces it, to the red haze of Titan and the organic molecules that lie behind it. All that is a range of extraordinary proportions, even given the breadth of the astronomical calling. Carl's work has not simply impressed his colleagues, but has guided and challenged them over so many years, and it has been recognized by honors that have poured in from across the world.

Carl Sagan is a master researcher and professional leader in his field, editing the leading planetary science journal, *Icarus*, for twelve years with great distinction. It is from that, and from the zest behind that, that Carl's teaching flows. I can give no more powerful tribute to the effectiveness of his teaching than the three marvelous tributes that you heard from students tonight. It is true that Carl's topic is the cosmos and his classroom is the world. Five hundred million people in sixty countries have watched *Cosmos* on the Public Broadcasting System. The book of that title appeared for seventy weeks on *The New York Times* best-seller list. That is teaching, and it is not just popular flash and quick example. Behind Carl's teaching lies scholarship of the most substantial kind. It is not simply the easy view and the slick explanation that Carl has so triumphantly brought to the general public; it is a more profound understanding and a more directed interest in science itself.

John Slaughter, former head of the National Science Foundation, once said, "Research is to teaching as sin is to confession. Unless you participate in the former, you have nothing much to say in the latter." I make no comment on sin and confession, but Carl's research lies behind his teaching. In an age when Washington is obsessed with the utilitarian value of science, when 60% of the budget of the National Science Foundation is to be allocated for so-called strategic research, what Carl, the master teacher, has done is to show how this most distinctive human quality – the need to know, the need to comprehend and grasp and understand – drives and motivates everything we do.

Carl is a master teacher: here on the Cornell campus, to undergraduates; there in Niger and, in fact, across the globe. But Carl is more than that. Carl is an inspiring example of the engaged, global citizen – not remote, not isolated, but involved. We live in an age of reductionism, and powerful and essential as that is in science, it is a devastatingly unsuccessful approach to much of the rest of life. Real fulfillment in society and in individual life comes not from reductionism, not only from analysis, but also from synthesis. Carl is a master of synthesis, and he has used that skill to engage us as a society in some of the biggest issues of our time.

That is not a popular position for an academic. It is supposed commonly that we should be more circumspect and impartial than that. It is a risky business, and, of course, it has its dangers. Political correctness has shown us what a chilling effect proselytization can have in the classroom. But Carl's engagement is of a different kind. With the conscience of a humanist and the consummate skill of the scientist, he addresses the needs of the society in which we live, and we are the richer for it.

Carl, we celebrate your birthday because the age of sixty, as I can testify, is a time of liberation. I love the story of the man who was sitting in the back of a cab driving through Washington and passed by the National Archives Building. He saw the inscription chiseled in stone, "What's past is prologue," and he asked the cabbie what it meant. The cabbie said, "That's just bureaucratic talk. It really means, you ain't seen nothing yet!"

Sixty is that kind of age, and I want you to join with me tonight in standing and drinking a toast to Carl Sagan: honored scientist, devoted teacher, master expositor, cosmic guide, respected colleague, exemplary Cornellian, trusted friend. Carl, we salute you.

Index